Meshless Methods in Solid Mechanics

Meshless Methods in Solid Mechanics

Youping Chen, James D. Lee, and Azim Eskandarian

Springer

Youping Chen
School of Engineering and Applied Science
George Washington University
Washington, DC
USA
ypchen2@gwu.edu

James Lee
School of Engineering and Applied Science
George Washington University
Washington, DC
USA
jdlee@gwu.edu

Azim Eskandarian
School of Engineering and Applied Science
George Washington University
Washington, DC
USA
eska@gwu.edu

ISBN 978-1-4419-2148-2 e-ISBN 978-0-387-33368-7

Printed on acid-free paper.

Printed in the United States of America. (TB/MVY)

9 8 7 6 5 4 3 2 1

springer.com

To the memory of my father, Dinghua Chen: an unforgettable mathematician, poet, educator, and a kind person.

Youping Chen

To my family, for their understanding and support.

James D. Lee

To my family and my parents for all the hard work and sacrifices they made towards my education.

Azim Eskandarian

Acknowledgments

The motivation that led to the development of this text began by our extensive research in crashworthiness and nonlinear finite element methods (FEM) at The George Washington University for vehicle and barrier safety. Due to some limitations of FEM, we explored alternative approaches, particularly suited for inverse analysis or design purposes. The majority of this research was conducted for the Federal Highway Administration of the U.S. Department of Transportation (DOT).

We would like to thank our research sponsor Dr. Mort Oskard and the Federal Highway Administration of the USDOT for the opportunity to explore new solution techniques for their critical problems. Their vision, understanding, and flexibility allowed us to dedicate more time to meshless methods among other alternatives. Their desire to seek new computational techniques for a variety of structural and mechanics problems motivated the development of the codes for this text.

We also would like to extend our appreciations to our graduate students, Mr. Xiaowei Zeng, Ms. Yajie Lei, and Mr. Liming Xiong for carefully reviewing the manuscripts and suggesting corrections.

Preface

Finite element method has been the dominant technique in computational mechanics in the past decades, and it has made significant contributions to the developments in engineering and science. Nevertheless, FEM is not well suited to problems having severe mesh distortion owing to extremely large deformations of materials, encountering moving discontinuities such as crack propagation along arbitrary and complex paths, involving considerable meshings and remeshings in structural optimization problems, or having multidomain of influence in multiphenomena physical problems. It is impossible to completely overcome those mesh-related difficulties by a mesh-based method. The highly structured nature of finite element approximations imposes severe penalties in seeking the solutions of those problems.

Distinguishing with finite element, finite difference, and finite volume methods, meshless method discretizes the continuum body only with a set of nodal points and the approximation is constructed entirely in terms of nodes. There is no need of mesh or elements in this method. It does not possess the mesh-related difficulties and provides an approach with more flexibility in the applications in engineering and science.

The meshless method started to capture the interest of a broader community of researchers only several years ago, and now it becomes a growing and evolving field. It is showing that this is a very rich area to be explored, and has great promise for many very challenging computational problems. On the one hand, great advances of meshless methods have been achieved. On the other hand, there are many aspects of meshless methods that could benefit from improvements. A broader community of researchers can bring diverse skills and backgrounds to bear on the task of improving this method.

We were invited to give a series of lecture on meshless methods, equivalent to one-semester graduate course, to research scientists at USDOT, in Spring 2003. This book is mainly based on the lecture notes distributed in that DOT class.

The main objective of this book is to provide a textbook for graduate courses on the numerical analysis of solid mechanics. It can also be used as a reference book for engineers and scientists who are exploring the physical world through computer simulations. Emphasis of this book is given to the understanding of the physical and mathematical characteristics of the procedures of computational continuum mechanics. It naturally brings the essence, advantages, and challenging problems of meshless methods into the picture.

This book covers the fundamentals of continuum mechanics, the integral formulation of continuum problems, the basic concepts of FEM, and the methodologies and applications of various meshless methods. It also provides general and detailed procedures of meshless analysis of elastostatics, elastodynamics, nonlocal continuum mechanics, and plasticity with a large number of numerical examples. Some basic and important mathematical methods are included in the Appendixes. For the readers who want to gain knowledge through hands-on experience, two meshless computer programs, one for elastostatics and elastodynamics and the other for the analysis of crack growth in elastoplastic continuum, are posted on the book's page at www.springer.com/0-387-30736-2. User's manuals are included in Appendix D and Appendix E, respectively.

Contents

1 Introduction

Foundation of Physical Theories

Physics is correctly considered to be the foundation of the natural sciences. Yet physical theories almost never deal with real objects or phenomena in nature, just because even a piece of metal is too complicated to be described exactly. Instead, theorists work with models that capture the most important properties of real objects.

There exist two fundamental physical models that provide foundations for all physical theories concerning modeling the material behavior: (1) microscopic discrete atomistic models and (2) macroscopic continuum models. The entire physical science is based on these two models. In the range of nanometer or below, one would only see collections of discrete particles moving under the influence of their mutual interaction forces. Fundamentally, the motion is governed by the law of quantum mechanics, although classical mechanics is a very good approximation in many problems. Hence, the natural description of microscopic physics is in terms of many-body dynamics. Quantum mechanics, crystal dynamics, molecular dynamics, and statistical mechanics are basic atomistic descriptions.

On the other hand, the results of centuries of experimental work are usually formalized into well-structured continuum theories covering the field of macroscopic physics, with such subdivisions as solid mechanics, fluid mechanics, elasticity, thermodynamics, electromagnetism, and acoustics. The purpose of such theories is to inscribe all objects of our perception into the familiar framework of a four-dimensional space-time. We want to describe what happens in every point of space and at every instant of time. As an abstraction of our immediate perception, matter and energy are considered as a continuum in this framework. Therefore, the natural mathematical representation of physical quantities in this perspective is by means of continuous or piecewise-continuous functions of the space coordinates $\boldsymbol{x}$ and time t.

Although the underlying physical concept may be the same, the descriptions by the two kinds of models are thoroughly different, whereas the success of both models has been demonstrated and tested throughout the history of science in explaining and predicting various physical phenomena.

The computer modeling and simulation of physical phenomena are basically divided into two categories corresponding to the two kinds of models: *atomistic modeling* and *partial differential equation-based continuum modeling*.

Atomistic Modeling and Computation

In an atomistic modeling, an explicit interaction rule between all atoms or between all electrons in the simulated system must be established. The vast array of atomic scale simulation methods can be then classified into two main classes of modeling approaches depending on how they model physical phenomena in the simulated system: *quantum mechanical* and *classical*.

It is well known that quantum mechanical representation is in term of wave function, $\Psi(\mathbf{r}, t)$, and the basic equation in quantum mechanics is the Schrödinger equation

$$\left\{-\frac{\hbar^2}{2m}\nabla^2 + V(\mathbf{r}, t)\right\} \Psi(\mathbf{r}, t) = i\hbar\frac{\partial}{\partial t}\Psi(\mathbf{r}, t), \tag{1.1}$$

$$\left\{-\frac{\hbar^2}{2m}\nabla^2 + V(\mathbf{r})\right\} \Psi(\mathbf{r}) = E\Psi(\mathbf{r}). \tag{1.2}$$

Equations (1.1) and (1.2) are the time-dependent and time-independent Schrödinger equations, respectively, in which $\hbar$ is Planck's constant, m is mass, and $-\frac{\hbar^2}{2m}\nabla^2 + V(\mathbf{r}, t)$ is the Hamiltonian.

For pure two-body systems, like the hydrogen atom, it is possible to solve the Schrödinger equation analytically. For system with few electrons, such as helium, the "many-electron" problem can be solved more or less exactly. However, more general many-electron systems cannot be treated with such precision any more. The understanding of the cooperative behavior of many interacting electrons and ions requires dramatic approximations for decoupling the formidable many-body problem.

Quantum mechanical computation of electronic structure of materials, based on decoupling the many-body problem, rest essentially on *Density Functional Theory* (DFT) and, in a very minor part, on the *Hartree–Fock* (HF) method. Both of them were established decades ago: the HF method in the 1930s, and DFT in the 1960s. But the birth date of genuine ab initio simulations of the properties of materials is much more recent. This breakthrough was made possible by the development of very innovative methods, techniques, approximations, and algorithms.

DFT provides a rigorous way to decouple the electron–electron interaction. By reformulating the problem in terms of the *ground-state* electron density, Hohenberg, Kohn, and Sham showed that the many-electron equation could be replaced by an equivalent set of independent, one-electron Schrödinger equations (Kohn–Sham equation)

$$\left[\frac{-\hbar^2}{2m}\nabla^2 + \sum_I V_{\text{ion}}(\mathbf{r} - \boldsymbol{R}_I) + V_H(\mathbf{r}) + \mathrm{V}_{\text{xc}}(\mathbf{r})\right] \Psi_i(\mathbf{r}) = E_i\Psi_i(\mathbf{r}), \tag{1.3}$$

where Ψ_i is the wave function of electronic state i, $V_{\text{ion}}(\mathbf{r} - \boldsymbol{R}_I)$ is the static electron-ion potential, $V_H(\mathbf{r})$ is the Hartree potential of the electrons, and $V_{\text{xc}}(\mathbf{r})$ is the exchange-correlation potential.

DFT implementations are often called "total energy" methods, since properties related to cohesion, structure, elasticity, and lattice dynamics are accessed by computing the total ground-state energy in various circumstances. Energy derivatives, which define forces and stresses, can be evaluated by finite differences. It is generally observed that DFT predicts basic ground-state structural properties such as lattice constant and bulk modulus to within a few percent of the experimental values, with no parameter fitting. Since it generally scales as N^3, where N is the number of electrons, its main drawback is the computational cost, which severely limits the size of the system that can be simulated. For a simulation involving metal oxides, it is estimated that a system up to several hundred atoms can be solved by DFT.

HF methods also make the one-electron approximation, and assume that the electron wave functions can be described as a combination of antisymmetric orbitals, typically Gaussians or atomic orbitals. Most HF methods can treat the electron correlation effect properly, and they are generally highly reliable. The methods scale as N^4, and primarily used for the study of small molecules.

Tight binding (TB) approach significantly reduces the cost of solving single-electron Schrödinger equation by making several further approximations. First, a minimal basis set of valence atomic orbitals is used to expand the wave function for each of the N-electrons in the system, so that typically 10–20 times fewer basis functions are used in TB than that in DFT. Second, self-consistency in the eigensolution is neglected so that iterative solutions are no longer required. Last, all interactions are parameterized. It is still a quantum mechanical method, but much less expensive than DFT. However, it requires parameter fitting, and its predications are generally not as accurate as those of DFT or HF.

Classical atomic scale models do not explicitly take into account the role of electrons in determining the material properties. Instead, physical mechanisms are described solely in terms of interactions between atoms. The interaction energy is expressed by an interatomic potential energy function whose form is often derived empirically and their parameters are obtained through fitting to experimental measurements. Computational approaches include molecular dynamics (MD) simulation, Monte Carlo simulation, and others.

In an MD simulation, the dynamical trajectories of atoms governed by the forces predicted by interatomic potentials are followed as a function of time. Because of its computational efficiency, MD has been widely used in applications requiring the collective behavior of large numbers of atoms. State-of-the-art MD calculations on parallel computers have studied the dynamical properties of systems containing millions of atoms. This would allow a specimen size up to microns on the time scale of picoseconds. Atomic scale modeling of biological or chemical system, mechanical phenomena, such as fracture and plasticity, are typically limited to the MD approaches due to the relatively large physical dimensions required. The ratio of computational cost of MD, TB, and DFT is roughly 1, 10^2–10^3 and 10^5.

An important shortcoming of classical atomic scale models is that they cannot describe the electronic structure of materials that results from electron interactions. In contrast, quantum mechanical models explicitly treat electron interactions by solving the governing equation of quantum mechanics. By starting with a fundamental description of materials, quantum mechanical approaches avoid the need to introduce empirical formulations and to numerically fit parameters.

PDE-Based Continuum Modeling and Computation

The macroscopic continuum physics are represented by a boundary value problem, a set of partial differential equations (PDE) with boundary conditions, in the continuum field. However, continuous problems can only be solved exactly by mathematical manipulation, and the available mathematical techniques usually limit the possibilities to oversimplified situations.

To overcome the intractability of the realistic type of continuum problem, various methods of *discretization* have been proposed and developed. All involve an *approximation* that approaches the true continuum solution as the number of discrete variables increases. There are various methods in this category. The purpose of those methods is to numerically solve the partial differential equations. Among those well established methods, there are finite difference (FD) methods, finite element methods, boundary element (BE) methods; and now we have meshless methods.

The FD method is an earliest classical numerical treatment for solving partial differential equations. In the FD method, we replace the continuous solution domain by a discrete set of lattice points. In each lattice point, we replace any differential operators by FD operators. By substituting the difference formulae into the PDE, a set of difference equations are obtained, which can then be easily solved.

Equations in differential forms can often be transformed into integral forms. The BE method utilizes this fact by transforming the differential operator defined in the domain to integral operators defined on the boundary. Hence, in the BE method only the boundary of the domain of interest requires discretization. As a consequence, the mesh generation is considerably simpler than other methods. For example, if the domain is either the interior or the exterior to a sphere then only the surface is divided into elements. The development of the BE method requires the governing PDE to be reformulated as an integral equation. Boundary solutions are obtained directly by solving the set of linear equations.

In the FE method, the solution domain can be discretized into a number of uniform or nonuniform finite elements that are connected via nodes. The change of the dependent variable with regard to location is approximated within each element by a shape function. The shape function is defined relative to the values of the variable at the nodes associated with each element. The original boundary value problem is then replaced with an equivalent integral formulation. The interpolation functions are then substituted into the integral equation, integrated, and combined with the results from all other elements in the solution domain. The results of this procedure can be reformulated into a matrix equation of the form,

$$\boldsymbol{M}\ddot{\boldsymbol{u}} + \boldsymbol{C}\dot{\boldsymbol{u}} + \boldsymbol{K}\boldsymbol{u} = \boldsymbol{F}, \tag{1.4}$$

where, usually, $\boldsymbol{M}$ is the mass matrix, $\boldsymbol{C}$ the damping matrix, $\boldsymbol{K}$ the stiffness matrix, $\boldsymbol{F}$ the applied force vector, and $\boldsymbol{u}$ the displacement field and the unknown variable.

Both the FE and FD methods are similar as the entire solution domain needs to be discretized, and a mesh is needed, while in the BE method only the bounding surfaces need to be meshed. For regular domains, the FD method is generally the easiest method to code and implement, but it usually requires special modifications to define irregular boundaries, abrupt changes in material properties, and complex boundary conditions. The FD method has the merit of computational simplicity and also often has an accuracy loss. The BE method has advantage where only the boundary solution is of interest, whereas FE method is superior to the BE method for representing nonlinearity and anisotropy.

Although FE and BE are alternatives for certain engineering problems, the FE method has been the dominant technique in computational mechanics in past decades, and it has made significant contributions to the advance in engineering and science.

In recent years, there has been considerable interest shown for *meshless method* in which there is no need of element or mesh. It discretizes the continuum body only with a set of nodal points and the approximation is constructed entirely in terms of nodes. The method is thus less susceptible to mesh distortion difficulties than the FE method. For a variety of problems with extremely large deformation, moving boundary discontinuities, or in optimization problems where re-meshing may be required, meshless methods are very attractive. The method has the promise to provide an approach with more flexibility in the applications in engineering and science.

2
Fundamental of Continuum Mechanics

Continuum mechanics is a branch of physical sciences concerned with the deformations and motions of continuous material media under the influence of external effects. External effects that influence bodies appear in the form of forces, displacements, and velocities that result from contact with other bodies, gravitational forces, thermal changes, chemical interactions, electromagnetic effects, and other environmental changes.

The theory of continuous media is built upon two strong foundations: (1) the basic laws of motion and (2) a constitutive theory. The basic laws of motion are the fundamental axioms of motion that are valid for all bodies irrespective of their constitution. They are the results of our experience with the physical world. The constitutive relations are constructed to take the nature of different materials into consideration. These relations also depend on the range of physical effects, which we wish to describe. Certain axioms are employed in the construction and restriction of the constitutive relations. The resulting equations nevertheless contain some unknown material parameters that must be determined through experiments and/or statistical mechanical considerations.

This chapter is a brief review of continuum mechanics devoted to a study of the basic laws of motions and the constitutive theory (cf. Eringen, 1989).

Kinematics

The purpose of this section is to study the local geometric changes and the motion of points in the continua. The relationship between the initial position of any material point in a body and its subsequent places is essential in the description of the local length, angle, and volume changes and translations and rotations of elements of the body. In this section, we are concerned with such changes and their measures both in space and in time, irrespective of the type of substance and the external effects.

Appendixes A and B provide the prerequisite explanation of all notations used in this section.

The material points of a continuous medium at $t = 0$ (undeformed state) occupy a region B that consists of the material volume V and its surface S. The position of a material point P in this region is expressed by $\boldsymbol{X}$ (or X_K, $K = 1, 2, 3$) in the material or Lagrangian coordinate system. After deformation takes place, at time t (deformed state), the material points of B occupy a region b consisting of a spatial volume v and its surface s. The position of the material point P at time t, denoted by p, is expressed by $\boldsymbol{x}$ (or x_k, $k = 1, 2, 3$) in the spatial or Eulerian coordinate system. In the description of the deformation and motion of a continuous medium, the use of two sets of coordinate systems (Lagrangian and Eulerian) makes many subtle points clearly understood. The motion of the material body that carries various material points through various spatial positions can now be expressed as

$$x_k = x_k(X_K, t) \text{ or } \boldsymbol{x} = \boldsymbol{x}(\boldsymbol{X}, t), \tag{2.1}$$

or conversely

$$X_K = X_K(x_k, t) \text{ or } \boldsymbol{X} = \boldsymbol{X}(\boldsymbol{x}, t). \tag{2.2}$$

It is noted that the motion, Eq. (2.1), takes a material point $\boldsymbol{X}$ in B at $t = 0$ to a spatial position $\boldsymbol{x}$ in b at time t and the inverse motion, Eq. (2.2), traces the material point occupying the spatial position $\boldsymbol{x}$ at time t back to its original position $\boldsymbol{X}$ at $t = 0$. Here, we take the assumption, known as the axiom of continuity, that the matter is indestructible and impenetrable. This is equivalent to the statement that the mappings, Eqs. (2.1) and (2.2), are single valued and possess continuous partial derivatives with respect to their arguments for whatever order as needed, except possibly at some singular points, curves, and surfaces. In other words, the motion and the inverse motion are the unique inverses of each other. It should be noted that, in practice, there are cases in which the axiom of continuity is violated. For example, material may have propagating cracks, may be broken into pieces, or may transmit shock waves and other types of discontinuities. Special attention and treatments must be given to those cases. This axiom is secured by the nonvanishing Jacobian, i.e.,

$$j \equiv \det(\nabla_X \boldsymbol{x}) = \left| \frac{\partial x_k}{\partial X_K} \right| \neq 0. \tag{2.3}$$

Now the notion of deformation gradients is introduced as

$$x_{k,K} \equiv \frac{\partial x_k}{\partial X_K}, \tag{2.4}$$

$$X_{K,k} \equiv \frac{\partial X_K}{\partial x_k}, \tag{2.5}$$

and from now on majuscules (minuscules) indices after a comma indicate partial differentiation with respect to Lagrangian (Eulerian) coordinates. The readers may verify the following identities as an exercise:

$$X_{K,k} = \frac{1}{2j} e_{KLM} e_{klm} x_{l,L} x_{m,M}, \tag{2.6}$$

$$j = \frac{1}{6} e_{KLM} e_{klm} x_{k,K} x_{l,L} x_{m,M}, \tag{2.7}$$

$$(j X_{K,k})_{,K} = 0, \tag{2.8}$$

$$(j^{-1} x_{k,K})_{,k} = 0, \tag{2.9}$$

$$\frac{\partial j}{\partial x_{k,K}} = j X_{K,k}. \tag{2.10}$$

The Green deformation tensor is defined as

$$C_{KL} \equiv x_{k,K} x_{k,L}. \tag{2.11}$$

When there is no deformation, the deformation gradient is reduced to a constant matrix, i.e.,

$$x_{k,K} \to \delta_{kK}, \tag{2.12}$$

where δ_{kK} is the matrix of direction cosines between the Lagrangian and Eulerian coordinates, which may be referred to as shifter. Then the Green deformation tensor is reduced to a Kronecker delta, δ_{KL}, in case of deformation free, hence it is natural to define the Lagrangian strain tensor as

$$E_{KL} \equiv (C_{KL} - \delta_{KL})/2. \tag{2.13}$$

Note that the shifter and Kronecker delta have completely different meanings.

Similarly, the Cauchy deformation tensor and the Eulerian strain tensor are defined as

$$c_{kl} \equiv X_{K,k} X_{K,l}, \tag{2.14}$$

$$\varepsilon_{kl} \equiv (\delta_{kl} - c_{kl})/2. \tag{2.15}$$

The physical meaning of the strain tensors may be seen as follows: the squares of the differential lengths in the deformed and undeformed states can be expressed, respectively, as

$$\mathrm{d}S^2 = \mathrm{d}X_K \mathrm{d}X_K = X_{K,k} \mathrm{d}x_k X_{K,l} \mathrm{d}x_l = c_{kl} \mathrm{d}x_k \mathrm{d}x_l, \tag{2.16}$$

$$\mathrm{d}s^2 = \mathrm{d}x_k \mathrm{d}x_k = x_{k,K} \mathrm{d}X_K x_{k,L} \mathrm{d}X_L = C_{KL} \mathrm{d}X_K \mathrm{d}X_L. \tag{2.17}$$

The difference $\mathrm{d}s^2 - \mathrm{d}S^2$ is the measure of the change of length due to deformation, and it is equal to

$$\begin{aligned} \mathrm{d}s^2 - \mathrm{d}S^2 &= (C_{KL} - \delta_{KL}) \mathrm{d}X_K \mathrm{d}X_L = (\delta_{kl} - c_{kl})\, \mathrm{d}x_k \mathrm{d}x_l \\ &= 2E_{KL} \mathrm{d}X_K \mathrm{d}X_L = 2\varepsilon_{kl} \mathrm{d}x_k \mathrm{d}x_l. \end{aligned} \tag{2.18}$$

Readers may find the measures of the changes of volume and area (cf. Problems of this chapter).

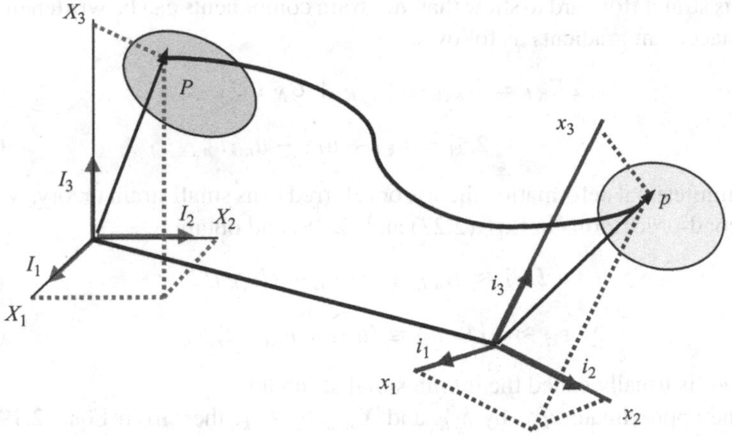

FIGURE 2.1. Motion and deformation of a point in a continuum.

One may verify that the Lagrangian and Eulerian strains are related as

$$E_{KL} = \varepsilon_{kl} x_{k,K} x_{l,L}, \tag{2.19}$$

$$\varepsilon_{kl} = E_{KL} X_{K,k} X_{L,l}. \tag{2.20}$$

We now define the displacement of a material point as the vector that extends from $\boldsymbol{X}$ in the undeformed state to $\mathbf{x}$ in the deformed state, i.e.,

$$\boldsymbol{u} \equiv \boldsymbol{x} - \boldsymbol{X} + \boldsymbol{b} = x_k \boldsymbol{i}_k - X_K \boldsymbol{I}_K + \boldsymbol{b}, \tag{2.21}$$

where $\boldsymbol{I}_K (K = 1, 2, 3)$ and $\boldsymbol{i}_k (k = 1, 2, 3)$ are the unit base vectors in the Lagrangian and Eulerian coordinate systems, respectively; $\boldsymbol{b}$ is the vector extending from the origin of the Lagrangian coordinate to the Eulerian coordinate (cf. Fig. 2.1). Then one may express a generic vector in either Lagrangian or Eulerian coordinate system as

$$\boldsymbol{u} = u_k \boldsymbol{i}_k = U_K \boldsymbol{I}_K, \quad \boldsymbol{b} = b_k \boldsymbol{i}_k = B_K \boldsymbol{I}_K. \tag{2.22}$$

Note that $\boldsymbol{I}_K \cdot \boldsymbol{I}_L = \delta_{KL}$, $\boldsymbol{i}_k \cdot \boldsymbol{i}_l = \delta_{kl}$, and $\boldsymbol{i}_k \cdot \boldsymbol{I}_K = \boldsymbol{I}_K \cdot \boldsymbol{i}_k = \delta_{kK} = \delta_{Kk}$. The shifter and the Kronecker have the following properties:

$$U_K = \delta_{Kk} u_k, \quad u_k = \delta_{kK} U_K, \tag{2.23}$$

$$\delta_{Kk} \delta_{Lk} = \delta_{KL}, \quad \delta_{Kk} \delta_{Kl} = \delta_{kl}. \tag{2.24}$$

It is worthwhile to note that any pair of vectors (tensors), that can be related in the form as in Eq. (2.23), is actually one vector (tensor) physically but represented in two-coordinate systems.

From Eq. (2.21) we may obtain

$$u_k = x_k - \delta_{Lk} X_L + b_k, \tag{2.25}$$

$$U_K = \delta_{Kk} x_k - X_K + B_K, \tag{2.26}$$

then it is straightforward to show that the strain components can be written in terms of displacement gradients as follows:

$$2E_{KL} = U_{K,L} + U_{L,K} + U_{N,K}U_{N,L}, \tag{2.27}$$

$$2\varepsilon_{kl} = u_{k,l} + u_{l,k} - u_{n,k}u_{n,l}. \tag{2.28}$$

In infinitesimal deformation theory, or referred to as small strain theory, we drop the second-order terms in Eqs. (2.27) and (2.28) and obtain

$$E_{KL} \approx \tilde{E}_{KL} \equiv (U_{K,L} + U_{L,K})/2, \tag{2.29}$$

$$\varepsilon_{kl} \approx \tilde{\varepsilon}_{kl} = e_{kl} \equiv (u_{k,l} + u_{l,k})/2, \tag{2.30}$$

where e_{kl} is usually called the infinitesimal strain tensor.

If one approximate $x_{k,K}$ by δ_{kK} and $X_{K,k}$ by δ_{Kk}, then from Eqs. (2.19) and (2.20), it results

$$\tilde{E}_{KL} = \tilde{\varepsilon}_{kl}\delta_{kK}\delta_{lL}, \tag{2.31}$$

$$\tilde{\varepsilon}_{kl} = \tilde{E}_{KL}\delta_{Kk}\delta_{Ll}, \tag{2.32}$$

which means in small strain theory the distinction between the Lagrangian strain and the Eulerian strain disappears.

The material time rate of change of a generic tensor $\boldsymbol{f}$ is defined as

$$\dot{f} \equiv \frac{\mathrm{d}f}{\mathrm{d}t} \equiv \left.\frac{\partial f}{\partial t}\right|_{X}, \tag{2.33}$$

where the subscript $\boldsymbol{X}$ accompanying the vertical bar indicates that $\boldsymbol{X}$ is held constant in the differentiation of $\boldsymbol{f}$. If $\boldsymbol{f}$ is a material function, i.e., $\boldsymbol{f} = \boldsymbol{f}(\boldsymbol{X}, t)$, then

$$\dot{f} \equiv \frac{\mathrm{d}f}{\mathrm{d}t} = \frac{\partial f(\boldsymbol{X}, t)}{\partial t}, \tag{2.34}$$

and if, on the other hand, $\boldsymbol{f}$ is a spatial function, i.e., $\boldsymbol{f} = \boldsymbol{f}(\boldsymbol{x}, t)$, then

$$\frac{\mathrm{d}f}{\mathrm{d}t} = \frac{\partial f(\boldsymbol{x}, t)}{\partial t} + \left.\frac{\partial f}{\partial x_k}\frac{\mathrm{d}x_k}{\mathrm{d}t}\right|_{X}. \tag{2.35}$$

The velocity vector $\boldsymbol{v}$ is defined as the material time derivative of the position vector of a material point. Note that, for a given material point, the Lagrangian coordinate $\boldsymbol{X}$ is a constant and the Eulerian coordinate $\boldsymbol{x}$ is a function of time. Now the material time derivative of a spatial function $\boldsymbol{f}$ can be written as

$$\dot{f} = \frac{\partial f}{\partial t} + f_{,k}v_k. \tag{2.36}$$

The acceleration vector $\boldsymbol{a}$ is the material time derivative of the velocity $\boldsymbol{v}$; therefore it can be expressed as

$$\dot{v}_k = \frac{\partial v_k}{\partial t} + v_{k,l}v_l, \tag{2.37}$$

where $v_{k,l}v_l$ is the convective term. It is straightforward to obtain the following frequently used expressions:

$$\frac{\mathrm{d}}{\mathrm{d}t}(dx_k) = v_{k,l}\mathrm{d}x_l, \tag{2.38}$$

$$\frac{\mathrm{d}}{\mathrm{d}t}(x_{k,K}) = v_{k,l}x_{l,K}, \tag{2.39}$$

$$\frac{\mathrm{d}}{\mathrm{d}t}(X_{K,k}) = -v_{l,k}X_{K,l}, \tag{2.40}$$

$$\frac{\mathrm{d}j}{\mathrm{d}t} = \frac{\mathrm{d}}{\mathrm{d}t}|x_{k,K}| = jv_{k,k}, \tag{2.41}$$

$$\frac{\mathrm{d}}{\mathrm{d}t}(\mathrm{d}s^2) = 2d_{kl}\mathrm{d}x_k\mathrm{d}x_l, \tag{2.42}$$

$$\dot{E}_{KL} = \frac{1}{2}\dot{C}_{KL} = d_{kl}x_{k,K}x_{l,L}, \tag{2.43}$$

where the deformation rate tensor $\boldsymbol{d}$ is defined as

$$d_{kl} = \frac{1}{2}(v_{k,l} + v_{l,k}). \tag{2.44}$$

Thus, we have shown that the displacements, strains, changes of length and volume, velocity, strain rates, deformation rate, etc., can all be derived from the motion, Eq. (2.1).

Basic Laws of Motion

In mechanics, for each material body there is an associated measure called *mass*. It is nonnegative, additive, and invariant under the motion. If the mass is absolutely continuous in the space variables, then there exists a mass density ρ so that the total mass of the body is determined by

$$M = \int_v \rho \mathrm{d}v, \tag{2.45}$$

where v is the volume of the region that the material body occupies. Now we define the *linear momentum* $\boldsymbol{P}$, the *angular momentum* $\boldsymbol{H}$ about the origin of the coordinate system, and the kinetic energy $\boldsymbol{K}$ of a continuous mass medium contained in v as

$$\boldsymbol{P} \equiv \int_v \rho \boldsymbol{v} \mathrm{d}v, \tag{2.46}$$

$$\boldsymbol{H} \equiv \int_v \boldsymbol{x} \times \rho \boldsymbol{v} \mathrm{d}v, \tag{2.47}$$

$$\boldsymbol{K} \equiv \frac{1}{2}\int_v \rho \boldsymbol{v} \cdot \boldsymbol{v} \mathrm{d}v. \tag{2.48}$$

Now we can enunciate four balance laws in continuum mechanics as follows:

Principle of Conservation of Mass. The total mass of the body is unchanged during the process of deformation and motion.

The global mass conservation law can be expressed as:

$$\int_v \rho \mathrm{d}v = \int_V \rho^0 \mathrm{d}V, \tag{2.49}$$

or

$$\frac{\mathrm{d}}{\mathrm{d}t}\int_v \rho \mathrm{d}v = 0. \tag{2.50}$$

Since $\int_v \rho \mathrm{d}v = \int_V \rho j \mathrm{d}V$, Eqs. (2.49) and (2.50), respectively, lead to

$$\int_V (\rho^0 - \rho j)\mathrm{d}V = 0, \tag{2.51}$$

$$\int_v (\dot{\rho} + \rho v_{k,k}) j \mathrm{d}v = \int_V (\dot{\rho} + \rho v_{k,k})\mathrm{d}V. \tag{2.52}$$

In this book, we further require that the conservation laws, such as Eqs. (2.51) and (2.52), to be valid for any arbitrary portion of the material body. This requirement can be satisfied if and only if the integrand is identically zero at any point in the volume of the material body, i.e.,

$$\rho^0 = \rho j, \quad \dot{\rho} + \rho v_{k,k} = 0, \tag{2.53}$$

which is the local form of the conservation law of mass.

Principle of Balance of Linear Momentum. The time rate of change of linear momentum is equal to the resultant force **F** acting on the body.

It can be expressed as

$$\frac{\mathrm{d}}{\mathrm{d}t}\int_v \rho \mathbf{v} \mathrm{d}v = \mathbf{F}. \tag{2.54}$$

Its local form can be written as

$$\rho \dot{v}_k = f_k. \tag{2.55}$$

Principle of Balance of Angular Momentum. The time rate of change of the angular momentum about any fixed point ***O*** is equal to the resultant moment **M** about that point.

It can be expressed as

$$\frac{\mathrm{d}}{\mathrm{d}t}\int_v \mathbf{x} \times \rho \mathbf{v} \mathrm{d}v = \mathbf{M}. \tag{2.56}$$

The local form can be obtained as

$$\rho e_{kij} x_i \dot{v}_j = m_k. \tag{2.57}$$

Principle of Conservation of Energy. The time rate of change of the kinetic energy plus the internal energy is equal to the sum of the rate of work done W by the external forces plus all other energy rates that enter or leave the material body.

The global form of conservation of energy can be written as

$$\frac{\mathrm{d}}{\mathrm{d}t}\int_v \rho\left(\frac{1}{2}v_k v_k + e\right)\mathrm{d}v = W + \sum_\alpha U_\alpha, \tag{2.58}$$

where U_α is the αth kind of energy that enters the body per unit time, which may be the heat energy, the electromagnetic energy, or the chemical energy. This principle implies that energies are additive, and if proper accounting is made due to all the external effects, what is left over to be balanced is the rate of the internal energy.

In analyzing the forces that act on the volume v of the material body, it is necessary to take into account the two types of forces:

1. Body force, which is the force proportional to the mass contained in v.
2. Surface force, which acts on the enclosing surface s of the volume v.

The body force can simply be expressed as $\int_v \rho \boldsymbol{f}\mathrm{d}v$, and $\boldsymbol{f}$ is the body force density (body force per unit mass). For surface force, we have to introduce the concepts of stress vector and stress tensor. Imagine a closed surface s within a material body. We would like to know the interaction between the material exterior to this surface and that within the surface. In this consideration, there arises the basic defining concept in continuum mechanics: the stress principle of Euler and Cauchy. Let Δs be a small surface element on s and $\boldsymbol{n}$ be a unit normal to Δs, with its direction outward from the interior of s. Then we can distinguish the positive and negative sides of Δs according to the direction of $\boldsymbol{n}$. Consider that the material lying on the positive side of the normal exerts a force $\Delta \boldsymbol{F}$ on the material lying on the negative side. The force $\Delta \boldsymbol{F}$ is a function of area and orientation of the surface. Assume that as Δs tends to be zero, the ratio $\Delta \boldsymbol{F}/\Delta s$ tends to be a definite limit, i.e.,

$$\lim_{\Delta s\to 0}\frac{\Delta \boldsymbol{F}}{\Delta s} = \frac{\mathrm{d}\boldsymbol{F}}{\mathrm{d}s} \equiv \boldsymbol{T}^n, \tag{2.59}$$

which is called the stress vector or the surface traction. In this book, it is also assumed that the moment of the forces acting on the surface Δs about any point within the area vanishes in the limit. Consider a special surface element Δs_k whose normal is along the x_k axis. Then $\boldsymbol{T}^k$ has three components T_1^k, T_2^k, and T_3^k acting along the direction of the coordinate axes x_1, x_2, and x_3, respectively. The index i of T_i^k refers to the direction of the force and the symbol k indicates the surface on which the force acts. Now we can construct a stress tensor $\boldsymbol{t}$ by linking its components to those in $\boldsymbol{T}^k$

$$t_{ij} = T_j^i. \tag{2.60}$$

One may verify Cauchy's formula that says knowing the nine components t_{ij} the stress vector $\boldsymbol{T}^n$ acting on any surface with a unit outward normal $\boldsymbol{n}$, with components n_1, n_2, n_3, can be represented by

$$T_i^n = t_{ji} n_j. \tag{2.61}$$

Therefore, the stress tensor is often referred to as the Cauchy stress.

Now the balance law of linear momentum can be rewritten as

$$\int_v \rho \dot{v}_k \mathrm{d}v = \int_v \rho f_k \mathrm{d}v + \oint_s t_{lk} n_l \mathrm{d}s. \tag{2.62}$$

Using the Green–Gauss theorem to convert the surface integral in Eq. (2.62) to the volume integral, it leads to the following balance law of linear momentum in local form

$$t_{lk,l} + \rho f_l = \rho \dot{v}_l. \tag{2.63}$$

The global form of the balance law of angular momentum can now be written as

$$\int_v e_{kij} x_i \dot{v}_j \mathrm{d}v = \int_v e_{kij} x_i f_j \mathrm{d}v + \oint_s e_{kij} x_i t_{mj} n_m \mathrm{d}s, \tag{2.64}$$

which, upon using Eq. (2.63) and the Green–Gauss theorem, leads to

$$e_{kij} t_{ij} = 0. \tag{2.65}$$

This simply means the stress tensor is symmetric, i.e.,

$$t_{ij} = t_{ji}. \tag{2.66}$$

It is worthwhile to mention that the symmetry of the stress tensor is the consequence of the balance law of angular momentum with the implicit assumptions that (1) the angular momentum is only the moment of the linear momentum, i.e., no spin inertia is incorporated, (2) the moment stress considered is only the moment of the Cauchy stress, and (3) no body couple is incorporated.

For conservation of energy, the rate of work done by the external forces can now be expressed as

$$W = \int_v \rho \boldsymbol{f} \cdot \boldsymbol{v} \mathrm{d}v + \oint_s \boldsymbol{T}^n \cdot \boldsymbol{v} \mathrm{d}s = \int \rho f_k v_k \mathrm{d}v + \oint_s t_{lk} v_k n_l \mathrm{d}s. \tag{2.67}$$

In this book, the nonmechanical energy considered is the thermal energy that can be written as

$$Q = \int_v \rho h \mathrm{d}v - \oint_s q_k n_k \mathrm{d}s, \tag{2.68}$$

where h denotes the distributed heat source having the dimension of energy per unit time per unit mass and $\boldsymbol{q}$ is the heat flux vector with the dimension of energy per unit time per unit area. Note that the surface integral in Eq. (2.68) $\oint_s \boldsymbol{q} \cdot \boldsymbol{n}\, \mathrm{d}s$ is the amount of heat energy flowing out per unit time through the enclosing surface s and that is why there is a minus sign associated with this surface integral. Interested readers should take a note that different authors may choose to use different sign convention for heat flux. After straightforward derivation, the local form of the conservation law of energy is obtained as

$$\rho \dot{e} = t_{ji} v_{i,j} - q_{j,j} + \rho h. \tag{2.69}$$

We may now formulate the *second law of thermodynamics*, may also be named as *principle of entropy*, as follows. There exist two scalar-valued functions of state, the absolute temperature θ and the entropy η, with the following properties:

1. Absolute temperature θ is strictly positive, which is a function of the empirical temperature only.
2. The entropy is additive, i.e., the entropy of the system is equal to the sum of entropies of its parts.
3. The entropy of a system can change in two distinct ways: by interaction with the surroundings and by changes taking place inside the system, i.e.,

$$\mathrm{d}S = \mathrm{d}S^{\mathrm{e}} + \mathrm{d}S^{\mathrm{i}}, \tag{2.70}$$

where the part of the increase due to the interaction with the surroundings $\mathrm{d}S^{\mathrm{e}}$ is equal to the heat absorbed by the system from its surroundings divided by the absolute temperature, i.e.,

$$\mathrm{d}S^{\mathrm{e}} = \frac{\mathrm{d}Q}{\theta}, \tag{2.71}$$

4. The change of entropy due to the changes taking place inside the system $\mathrm{d}S^{\mathrm{i}}$ is never negative, i.e.,

$$\mathrm{d}S^{\mathrm{i}} = \mathrm{d}S - \mathrm{d}Q/\theta \geq 0. \tag{2.72}$$

If $\mathrm{d}S^{\mathrm{i}}$ is zero, the process is said to be thermodynamically reversible. If $\mathrm{d}S^{\mathrm{i}}$ is positive, the process is said to be thermodynamically irreversible. The remaining case, $\mathrm{d}S^{\mathrm{i}} < 0$, never occurs in nature. Now we may enunciate the following fundamental law of entropy.

Principle of Entropy. The time rate of change of the total entropy is never less than the sum of the influx of entropy through the surface of the body and the entropy supplied by the body sources.

We may express the principle of entropy as an inequality

$$\frac{\mathrm{d}}{\mathrm{d}t}\int_v \rho\eta \mathrm{d}v + \oint_s \frac{q_k}{\theta} n_k \mathrm{d}s - \int_v \frac{\rho h}{\theta}\mathrm{d}v \geq 0. \tag{2.73}$$

The local form can be written as

$$\rho\dot{\eta} + \frac{q_{k,k}}{\theta} - \frac{q_k\theta_{,k}}{\theta^2} - \frac{\rho h}{\theta} \geq 0, \tag{2.74}$$

which is often called the Clausius–Duhem (CD) inequality. It is important to familiarize ourselves with the concept that absolute temperature and entropy are attributes to a material body, just as its mass or its electric charges are. We will not attempt to define them in terms of other quantities regarded as simpler. The justification of those postulates that form the basis of the classical thermodynamics is the empirical fact that all conclusions derived from these assumptions are without exception in agreement with the experimentally observed behavior of systems in nature at the macroscopic scale.

If the Helmholtz free energy function is defined as

$$\psi \equiv e - \eta\theta, \tag{2.75}$$

then the (CD) inequality can be rewritten as

$$-\rho(\dot{\psi} + \eta\dot{\theta}) + t_{ij}v_{j,i} - \frac{q_k\theta_{,k}}{\theta} \geq 0. \tag{2.76}$$

It is convenient to introduce the second-order Piola–Kirchhoff stress tensor as

$$T_{KL} \equiv j t_{kl} X_{K,k} X_{L,l}, \tag{2.77}$$

which enables one to express the Cauchy stress tensor in terms of the second-order Piola–Kirchhoff stress tensor as

$$t_{kl} = j^{-1} T_{KL} x_{k,K} x_{l,L}. \tag{2.77)*$$

After straightforward derivation, the balance law of linear momentum can be rewritten as

$$(T_{KL}x_{k,L})_{,K} + \rho^0(f_k - \dot{v}_k) = 0, \tag{2.78}$$

and also the symmetry of the Cauchy stress tensor implies the symmetry of the Piola–Kirchhoff stress tensor, i.e.,

$$T_{KL} = T_{LK}. \tag{2.79}$$

The conservation law of energy and the CD inequality can be rewritten in Lagrangian form, respectively, as

$$\rho^0\dot{e} = T_{KL}\dot{E}_{KL} - Q_{K,K} + \rho^0 h, \tag{2.80}$$

$$-\rho^0(\dot{\psi} + \eta\dot{\theta}) + T_{KL}\dot{E}_{KL} - Q_K\theta_{,K}/\theta \geq 0, \tag{2.81}$$

where

$$Q_K \equiv j X_{K,k} q_k. \tag{2.82}$$

Constitutive Theory

From the fundamental laws of continuum mechanics, including the CD inequality, we have eight equations: one each for conservation of mass and energy and three each for balance of linear and angular momenta; and 19 unknown variables, namely, ρ, v_k, t_{ij}, e, q_k, θ, and η. Mathematically speaking, 11 additional equations must be supplied to make the problems in continuum mechanics determinate. Physically, we understand that the fundamental laws are valid for all types of media irrespective of their internal constitutions. Obviously, material bodies with the same geometry, when subjected to identical external effects, respond differently. Internal constitution of matter is responsible for these different responses. In continuum mechanics, we are not concerned with the atomic structure and the interatomic forces of the matter; rather, we are interested in the macroscopic behaviors of the matter resulting from its internal constitution. To this end, therefore, we need equations, named constitutive equations, to describe the behaviors of each material in the range of phenomena to be covered. For example, if we are interested in solids, which upon the relief of applied loads return to their original undeformed states, then we are dealing with elastic solids. However, we also know from our daily experience that all solids undergo a permanent deformation when the applied loads exceed certain limits. To study the behavior of solids outside the elastic limits, we need the constitutive equations of plasticity. Thus, a set of constitutive equations is intended only to describe a range of physical phenomena decided at the outset for a given material.

For a constitutive theory to adequately represent a material, certain physical and mathematical requirements have to be satisfied. The following axioms are basic to the formulation of the constitutive equations (Eringen, 1989, 1999):

Axiom of Causality. We consider the motion and temperature of the material points of a body as self-evident observable effects in every thermomechanical behavior of the body. The remaining quantities (other than those derivable from the motion and temperature) that enter the expression of the CD inequality are the causes or the dependent constitutive variables.

The axiom of causality is intended for the selection of the independent and dependent constitutive variables. Since the motion $\boldsymbol{x} = \boldsymbol{x}(\boldsymbol{X}, t)$ and temperature $\theta = \theta(\boldsymbol{X}, t)$ are chosen as the "effects," it is seen that $x_{k,K}$, $\rho = \rho^0/\det(x_{k,K})$, v_k, $v_{k,l}$, d_{kl}, $\dot{E}_{KL}$, $\theta_{,k}$, and θ are all derivable from the motion and temperature. Then, from either Eq. (2.76) or Eq. (2.81), the dependent constitutive variables are ψ, η, t_{kl} or T_{KL}, q_k or Q_K.

Axiom of Determinism. The value of the thermomechanical constitutive functions $\{\psi, \eta, \boldsymbol{T}, \boldsymbol{Q}\}$ at a material point $\boldsymbol{X}$ of the body B at time t is determined by the histories of the motion and temperature of all material points in B.

This axiom is a principle of exclusion. It excludes the dependence of the material behavior at any point outside the body and any future events. Following this axiom one may write

$$\boldsymbol{T}(\boldsymbol{X}, t) = \hat{\boldsymbol{T}}\{\boldsymbol{x}(\boldsymbol{X}', t'), \theta(\boldsymbol{X}', t'), \boldsymbol{X}, t\}, \tag{2.83}$$

where $\hat{\boldsymbol{T}}$ is a tensor-valued functional defined over the field of functions $\boldsymbol{x}(\boldsymbol{X}', t')$ and $\theta(\boldsymbol{X}', t')$ with

$$\boldsymbol{X}' \in B \text{ and } t' \leq t, \tag{2.84}$$

and it is also a function of $\boldsymbol{X}$ and $\boldsymbol{t}$. Similarly, $\boldsymbol{Q}$, η, ψ can be written as functionals with the same list of arguments as in $\hat{\boldsymbol{T}}$.

Axiom of Equipresence. At the outset all constitutive functionals should be expressed in terms of the same list of independent constitutive variables until the contrary is deduced.

This axiom is a precautionary measure. It helps us not to be prejudiced against a certain class of variables and favor others in the expression of constitutive functionals. Later, the axiom of admissibility and various approximations may eliminate the dependence on some of these variables. Until such is shown to be the case, we should follow the axiom of equipresence.

Axiom of Neighborhood. The values of the independent constitutive variables at distant material points from $\boldsymbol{X}$ do not affect appreciably the values of the dependent constitutive variables at $\boldsymbol{X}$.

There is no single way to formulate this axiom. Here, we present a popular and useful formulation, named *smooth neighborhood*, for this book. The Taylor series expansion of $\boldsymbol{x}(\boldsymbol{X}', t')$ about $\boldsymbol{X}' = \boldsymbol{X}$ for all $t' \leq t$ is written as

$$\begin{aligned}\boldsymbol{x}(\boldsymbol{X}', t') &= \boldsymbol{x}(\boldsymbol{X}, t') + (X'_K - X_K)\boldsymbol{x}_{,K}(\boldsymbol{X}, t') \\ &\quad + \frac{1}{2!}(X'_K - X_K)(X'_L - X_L)\boldsymbol{x}_{,KL}(\boldsymbol{X}, t') \\ &\quad + \frac{1}{3!}(X'_K - X_K)(X'_L - X_L)(X'_M - X_M)\boldsymbol{x}_{,KLM}(\boldsymbol{X}, t') + \cdots .\end{aligned} \tag{2.85}$$

Similarly, one may expand $\theta(\boldsymbol{X}', t')$ into a Taylor series about $\boldsymbol{X}' = \boldsymbol{X}$ for all $t' \leq t$. If the constitutive functionals are sufficiently smooth so that they can be approximated by functionals in the field of real functions $\boldsymbol{x}(\boldsymbol{X}, t'), \boldsymbol{x}_{,K}(\boldsymbol{X}, t'), \boldsymbol{x}_{,KL}(\boldsymbol{X}, t'), \boldsymbol{x}_{,KLM}(\boldsymbol{X}, t'), \ldots$ and $\theta(\boldsymbol{X}, t'), \theta_{,K}(\boldsymbol{X}, t')$, $\theta_{,KL}(\boldsymbol{X}, t'), \theta_{,KLM}(\boldsymbol{X}, t'), \ldots$, the material is said to satisfy the smooth neighborhood hypothesis and may be named as material of gradient type.

Axiom of Memory. The values of the independent constitutive variables at distant past from the present do not affect appreciably the values of the dependent constitutive variables at the present time t.

This axiom is the counterpart of the axiom of neighborhood in the time domain. Accordingly, the memory of the past motions and temperatures of any material point decays rapidly. As in the case of axiom of neighborhood, no unique mathematical formulation can be made of this axiom. Here, we adopt the *smooth memory* hypothesis. The Taylor series expansions of the motion and temperature at $t' = t$ for all $\boldsymbol{X}'$ in B can be written as

$$\boldsymbol{x}(\boldsymbol{X}', t') = \boldsymbol{x}(\boldsymbol{X}', t) + (t' - t)\dot{\boldsymbol{x}}(\boldsymbol{X}', t) + \cdots + \tfrac{1}{n!}(t' - t)^n \overset{(n)}{\boldsymbol{x}}(\boldsymbol{X}', t) + \cdots, \tag{2.86}$$

$$\theta(\boldsymbol{X}', t') = \theta(\boldsymbol{X}', t) + (t' - t)\dot{\theta}(\boldsymbol{X}', t) + \cdots + \tfrac{1}{n!}(t' - t)^n \overset{(n)}{\theta}(\boldsymbol{X}', t) + \cdots, \tag{2.87}$$

where

$$\overset{(n)}{\boldsymbol{x}} \equiv \frac{\mathrm{d}^n \boldsymbol{x}}{\mathrm{d}t^n}, \ \overset{(n)}{\theta} \equiv \frac{\mathrm{d}^n \theta}{\mathrm{d}t^n}. \tag{2.88}$$

For materials that satisfy the smooth memory hypothesis, the constitutive functionals can be reduced to functionals in the field of real functions $\boldsymbol{x}(\boldsymbol{X}', t), \dot{\boldsymbol{x}}(\boldsymbol{X}', t), \ddot{\boldsymbol{x}}(\boldsymbol{X}', t), \ldots \overset{(n)}{\boldsymbol{x}}(\boldsymbol{X}', t), \ldots$ and $\theta(\boldsymbol{X}', t), \dot{\theta}(\boldsymbol{X}', t), \dot{\theta}(\boldsymbol{X}', t), \ldots \overset{(n)}{\theta}(\boldsymbol{X}', t), \ldots$, and these materials may be named as materials of rate type.

Axiom of Objectivity. The constitutive equations must be form-invariant with respect to rigid motions of the spatial frame of reference.

This is one of the very few important concepts in continuum mechanics, especially in the construction of constitutive equations. To understand *objectivity*, we first introduce two definitions as follows:

Definition 1. Two motions are called objectively equivalent if and if only

$$\bar{x}_k(X, \bar{t}) = Q_{kl}(t)x_l(X, t) + b_k(t), \quad \bar{t} = t - a, \tag{2.89}$$

where

$$Q_{km}(t)Q_{lm}(t) = Q_{mk}(t)Q_{ml}(t) = \delta_{kl}, \quad \det(Q_{kl}) = 1. \tag{2.90}$$

These two objectively equivalent motions differ only relative to the reference frame and time. For a fixed frame and time, the two motions can be made to coincide by the superposition of a rigid motion of one and by a shift of time.

Definition 2. Any tensorial quantity is said to be objective if in any two objectively equivalent motions it obeys the following tensor transformation law for all times

$$\bar{A}_{klm}\ldots(\boldsymbol{X}, \bar{t}) = Q_{kk'}(t)Q_{ll'}(t)Q_{mm'}(t)\ldots A_{k'l'm'}\ldots(\boldsymbol{X}, t). \tag{2.91}$$

According to the axiom of objectivity, for example, a second-order tensor-valued functional, such as the Cauchy stress tensor, should satisfy the following requirement:

$$\bar{\boldsymbol{t}}(\boldsymbol{X}, \bar{t}) = \boldsymbol{Q}(t)\bar{\boldsymbol{t}}(\boldsymbol{X}, t)\boldsymbol{Q}^t(t), \tag{2.92}$$

which implies

$$\begin{aligned} &\hat{t}_{ij}\{Q_{kl}(t')x_l(\boldsymbol{X}', t') + b_k(t'), \theta(\boldsymbol{X}', t'), \boldsymbol{X}, t - a\} \\ &\quad = Q_{im}(t)Q_{jn}(t)\hat{t}_{mn}\{x_k(\boldsymbol{X}', t'), \theta(\boldsymbol{X}', t'), \boldsymbol{X}, t\}. \end{aligned} \tag{2.93}$$

Axiom of Material Invariance. Constitutive equations must be form-invariant with respect to a group of orthogonal transformations $\{\boldsymbol{S}\}$ and translations $\{\boldsymbol{B}\}$ of the material coordinates, which reflects the symmetric conditions of the material body.

This axiom says that the constitutive functionals, e.g., stresses, should transform according to

$$\hat{\boldsymbol{t}}\{\boldsymbol{x}(\boldsymbol{X}', t'), \theta(\boldsymbol{X}', t'), \boldsymbol{X}, t\} = \hat{\boldsymbol{t}}\{\boldsymbol{x}(\boldsymbol{S}\boldsymbol{X}' + \boldsymbol{B}, t'), \theta(\boldsymbol{S}\boldsymbol{X}' + \boldsymbol{B}, t'), \boldsymbol{S}\boldsymbol{X} + \boldsymbol{B}, t\}, \tag{2.94}$$

under the transformation of the Lagrangian coordinate system

$$\overline{\boldsymbol{X}} = \boldsymbol{S}\boldsymbol{X} + \boldsymbol{B}, \tag{2.95}$$

where $\boldsymbol{S}$ is subjected to $\boldsymbol{S}\boldsymbol{S}^t = \boldsymbol{S}^t\boldsymbol{S} = \boldsymbol{I}$. It should be emphasized that if the material in question does not have any symmetry, then this axiom does not impose any constraint on the constitutive equations.

Axiom of Admissibility. All constitutive equations must be consistent with the basic principles of continuum mechanics, including the CD inequality.

Thermoviscoelastic Solid and Its Special Cases

In this section, we are going to formulate the constitutive equations for thermoviscoelastic solid. To begin with, let the constitutive functions be written as

$$\psi(\boldsymbol{X}, t) = \psi(\boldsymbol{E}, \dot{\boldsymbol{E}}, \theta, \dot{\theta}, \nabla\theta, \boldsymbol{X}), \tag{2.96}$$

$$\eta(\boldsymbol{X}, t) = \eta(\boldsymbol{E}, \dot{\boldsymbol{E}}, \theta, \dot{\theta}, \nabla\theta, \boldsymbol{X}), \tag{2.97}$$

$$\boldsymbol{T}(\boldsymbol{X}, t) = \boldsymbol{T}(\boldsymbol{E}, \dot{\boldsymbol{E}}, \theta, \dot{\theta}, \nabla\theta, \boldsymbol{X}), \tag{2.98}$$

$$\boldsymbol{Q}(\boldsymbol{X}, t) = \boldsymbol{Q}(\boldsymbol{E}, \dot{\boldsymbol{E}}, \theta, \dot{\theta}, \nabla\theta, \boldsymbol{X}), \tag{2.99}$$

which imply that this material is strain, strain rate, temperature, temperature rate, and temperature gradient dependent. It is recognized that the constitutive functions, Eqs. (2.96)–(2.99) follow the axioms of causality, determinism, equipresence, neighborhood, memory, and objectivity. Substituting Eqs. (2.96)–(2.99) into the entropy inequality, Eq. (2.81), one obtains

$$-\rho^0\left(\frac{\partial\psi}{\partial\boldsymbol{E}} : \dot{\boldsymbol{E}} + \frac{\partial\psi}{\partial\dot{\boldsymbol{E}}} : \ddot{\boldsymbol{E}} + \frac{\partial\psi}{\partial\theta}\dot{\theta} + \frac{\partial\psi}{\partial\dot{\theta}}\ddot{\theta} + \frac{\partial\psi}{\partial\nabla\theta}\cdot\nabla\dot{\theta}\right)$$
$$+\boldsymbol{T} : \dot{\boldsymbol{E}} - \frac{\boldsymbol{Q}\cdot\nabla\theta}{\theta} \geq 0. \tag{2.100}$$

It is seen that (2.100) is linear in $\ddot{\boldsymbol{E}}$, $\ddot{\theta}$, and $\nabla\dot{\theta}$. The necessary and sufficient conditions for the CD inequality to be valid are

$$\frac{\partial\psi}{\partial\dot{\boldsymbol{E}}} = \frac{\partial\psi}{\partial\dot{\theta}} = \frac{\partial\psi}{\partial\nabla\theta} = 0 \quad\Rightarrow\quad \psi = \psi(\boldsymbol{E}, \theta, \boldsymbol{X}), \tag{2.101}$$

$$-\rho^0\left(\frac{\partial\psi}{\partial\theta} + \eta\right)\dot{\theta} + \left(\boldsymbol{T} - \rho^0\frac{\partial\psi}{\partial\boldsymbol{E}}\right) : \dot{\boldsymbol{E}} - \frac{\boldsymbol{Q}\cdot\nabla\theta}{\theta} \geq 0. \tag{2.102}$$

Now we may separate the entropy and the Piola–Kirchhoff stresses into two parts: the reversible (elastic) part and the irreversible (dissipative) part, as

$$\eta = \eta^{\mathrm{e}} + \eta^{\mathrm{d}} = -\frac{\partial\psi}{\partial\theta} + \eta^{\mathrm{d}}(\boldsymbol{E}, \dot{\boldsymbol{E}}, \theta, \dot{\theta}, \nabla\theta, \boldsymbol{X}), \tag{2.103}$$

$$\boldsymbol{T} = \boldsymbol{T}^{\mathrm{e}} + \boldsymbol{T}^{\mathrm{d}} = \rho^0\frac{\partial\psi}{\partial\boldsymbol{E}} + \boldsymbol{T}^{\mathrm{d}}(\boldsymbol{E}, \dot{\boldsymbol{E}}, \theta, \dot{\theta}, \nabla\theta, \boldsymbol{X}). \tag{2.104}$$

Then the CD inequality is expressed as

$$-\rho^0\eta^{\mathrm{d}}\dot{\theta} + \boldsymbol{T}^{\mathrm{d}} : \dot{\boldsymbol{E}} - \frac{\boldsymbol{Q}\cdot\nabla\theta}{\theta} \geq 0. \tag{2.105}$$

It is seen, from Eq. (2.105), that there are three pairs of thermodynamic conjugates that contribute to the irreversibility of this material: the irreversible part of the entropy and the temperature rate, the irreversible part of the second-order Piola–Kirchhoff stress, and the Lagrangian strain rate, the heat flux and the temperature gradient.

Several special constitutive theories may be deduced as follows.

Simple Theory for Thermoviscoelastic Solid

For constitutive functions without temperature rate dependence (cf. Eqs. (2.96)–(2.99)), the dissipative part of the entropy vanishes, i.e.,

$$\eta = -\frac{\partial\psi}{\partial\theta}, \tag{2.106}$$

which means we recover the classical Gibbs equation for entropy; and the CD inequality is reduced to

$$\boldsymbol{T}^{\mathrm{d}} : \dot{\boldsymbol{E}} - \frac{\boldsymbol{Q}\cdot\nabla\theta}{\theta} \geq 0. \tag{2.107}$$

Viscoelasticity

If we exclude both temperature rate and gradient from the list of independent constitutive variables, the following equations are obtained

$$\eta = -\frac{\partial\psi}{\partial\theta}, \tag{2.108}$$

$$Q_K = 0, \tag{2.109}$$

$$\boldsymbol{T}^{\mathrm{d}} : \dot{\boldsymbol{E}} \geq 0. \tag{2.110}$$

It implies that the heat flux vanishes and the dissipation is only due to the viscous effect.

Thermoelasticity

If one excludes the strain rate and the temperature rate from the constitutive functions, it results in the following:

$$\eta^{\mathrm{d}} = \boldsymbol{T}^{\mathrm{d}} = 0, \tag{2.111}$$

$$Q_K\theta_{,K} \leq 0. \tag{2.112}$$

Equation (2.112) means the angle between the heat flux and the temperature gradient is greater than 90°, in other words, it says that the heat flows from the hot region to the cold region. This is a well-known physical phenomenon, but now we note that it is just a special case in continuum mechanics.

We further assume that the Helmholtz free energy density can be expanded as a polynomial in its arguments up to second order, i.e.,

$$\psi = \psi^0 - \eta^0 T - \gamma T^2/2T^0 - B_{KL}E_{KL}T/\rho^0 + A_{KLMN}E_{KL}E_{MN}/2\rho^0, \tag{2.113}$$

where $\psi^0, \eta^0, \gamma, B_{KL} = B_{LK}$ and $A_{KLMN} = A_{MNKL} = A_{LKMN} = A_{KLNM}$ are material constants, and they can be functions of Lagrangian coordinate $\mathbf{X}$ if there exists material inhomogeneity; T is the temperature variation, from the reference temperature T^0, having the following constraints

$$\theta \equiv T + T^0, \quad T^0 > 0, \quad |T| << T^0. \tag{2.114}$$

Then it is straightforward to obtain

$$\begin{aligned} \eta &= -\frac{\partial \psi}{\partial \theta} = \eta^0 + \gamma T/T^0 + B_{KL}E_{KL}/\rho^0, \\ T_{KL} &= \rho^0 \frac{\partial \psi}{\partial E_{KL}} = -B_{KL}T + A_{KLMN}E_{MN}. \end{aligned} \tag{2.115}$$

The heat flux Q_K is assumed to be linearly proportional to the temperature gradient, i.e.,

$$Q_K = -H_{KL}T_{,L}. \tag{2.116}$$

Following the Onsager's postulate, one may show that the $\mathbf{H}$ matrix in Eq. (2.116) is symmetric (cf. Eringen, 1999). It is seen that the CD inequality, Eq. (2.112), now implies

$$-H_{KL}T_{,L}T_{,K} \leq 0, \tag{2.117}$$

which means the material property matrix $\mathbf{H}$ has to be positive definite. Now the conservation law of energy, Eq. (2.80), can be expressed as

$$\rho^0 \gamma \frac{T + T^0}{T^0}\dot{T} + (T + T^0)B_{KL}\dot{E}_{KL} = (H_{KL}T_{,L})_{,K} + \rho^0 h. \tag{2.118}$$

Elasticity

For constitutive functions that only retain the strain, temperature, and the Lagrangian coordinate as the independent constitutive variables, we have

$$\eta = -\frac{\partial \psi}{\partial \theta}, \quad \boldsymbol{T} = \rho^0 \frac{\partial \psi}{\partial \boldsymbol{E}}, \quad \boldsymbol{Q} = 0; \tag{2.119}$$

and the CD inequality is reduced to a null statement $0 \geq 0$, which simply means there is no dissipation and the thermomechanical process is reversible. We further assume that $B_{KL} = 0$, then the equilibrium equation, Eq. (2.78), and the energy equation, Eq. (2.118), are reduced to

$$\{A_{KLMN}E_{MN}(\delta_{kL} + \delta_{kM}U_{M,L})\}_{,K} + \rho^0(f_k - \dot{v}_k) = 0, \tag{2.120}$$

$$\rho^0\gamma \tfrac{T+T^0}{T^0}\dot{T} = \rho^0 h, \tag{2.121}$$

which means the displacement field and the temperature field are completely decoupled.

Heat-Conducting Fluid

Definition. In continuum mechanics, a material body is called a fluid if every configuration of the body leaving the density unchanged can be taken as the reference configuration.

Following this definition, the independent constitutive variables appearing in Eqs. (2.96)–(2.99) for fluid are reduced to $\rho, d_{kl}, \theta, \dot{\theta}$, and $\theta_{,k}$; The constitutive equations are obtained as

$$\psi = \psi(\rho^{-1}, \theta), \tag{2.122}$$

$$\eta = -\frac{\partial\psi}{\partial\theta} + \eta^{\mathrm{d}}(\rho^{-1}, d_{kl}, \theta, \dot{\theta}, \theta_{,k}), \tag{2.123}$$

$$t_{ij} = -\frac{\partial\psi}{\partial\rho^{-1}}\delta_{ij} + t^{\mathrm{d}}_{ij}(\rho^{-1}, d_{kl}, \theta, \dot{\theta}, \theta_{,k}), \tag{2.124}$$

$$q_k = q_k(\rho^{-1}, d_{kl}, \theta, \dot{\theta}, \theta_{,k}), \tag{2.125}$$

$$\boldsymbol{t}^{\mathrm{d}} : \boldsymbol{d} - \rho\eta^{\mathrm{d}}\dot{\theta} - \tfrac{q_k\theta_{,k}}{\theta} \geq 0. \tag{2.126}$$

Equations (2.123)–(2.125) are saying that the irreversible parts of entropy and the Cauchy stress and the heat flux are isotropic scalar-valued, tensor-valued, and vector-valued functions of scalars $\rho^{-1}, \theta, \dot{\theta}$, vector $\nabla\theta$, and tensor $\boldsymbol{d}$. According to the representation theorems for isotropic functions (cf. Wang, 1970, 1971 and Appendixes A and B), η^{d}, $\boldsymbol{q}$, and $\boldsymbol{t}^{\mathrm{d}}$ can be written as

$$\eta^{\mathrm{d}} = \eta^{\mathrm{d}}(\rho^{-1}, \theta, \dot{\theta}, I_1, I_2, I_3, I_4, I_5, I_6), \tag{2.127}$$

$$q_k = a_1\theta_{,k} + a_2 d_{kl}\theta_{,l} + a_3 d_{kl}d_{lm}\theta_{,m}, \tag{2.128}$$

$$\begin{aligned} t^{\mathrm{d}}_{ij} &= b_1\delta_{ij} + b_2 d_{ij} + b_3 d_{ik}d_{kj} + b_4\theta_{,i}\theta_{,j} \\ &\quad + b_5(\theta_{,i}d_{jk}\theta_{,k} + d_{ik}\theta_{,k}\theta_{,j}) \\ &\quad + b_6(\theta_{,i}d_{jk}d_{kl}\theta_{,l} + d_{ik}d_{kl}\theta_{,l}\theta_{,j}). \end{aligned} \tag{2.129}$$

where

$$I_1 = d_{kk}, \tag{2.130}$$

$$I_2 = d_{kl}d_{lk}, \tag{2.131}$$

$$I_3 = d_{kl}d_{lm}d_{mk}, \tag{2.132}$$

$$I_4 = \theta_{,k}\theta_{,k}, \tag{2.133}$$

$$I_5 = \theta_{,k}d_{kl}\theta_{,l}, \tag{2.134}$$

$$I_6 = \theta_{,k}d_{kl}d_{lm}\theta_{,m}; \tag{2.135}$$

and $a_i(i = 1, 2, 3)$ and $b_j(j = 1, 2, \ldots, 6)$ are functions of ρ^{-1}, θ, $\dot{\theta}$, and $I_k(k = 1, 2, \ldots, 6)$. The CD inequality is now reduced to

$$\begin{aligned} b_1 I_1 + b_2 I_2 + b_3 I_3 + b_4 I_5 + 2b_5 I_6 + 2b_6 I_7 - \rho\eta^{\mathrm{d}}\dot{\theta} \\ -\frac{1}{\theta}(a_1 I_4 + a_2 I_5 + a_3 I_6) \geq 0, \end{aligned} \tag{2.136}$$

where $I_7 \equiv d^3_{ij}\theta_{,i}\theta_{,j}$.

In summary, we have derived the balance laws for mass, linear momentum, angular momentum, energy, and entropy (Eqs. (2.53), (2.63), (2.66), (2.69), and (2.74)), the constitutive equations for the general thermoviscoelastic solid (Eqs. (2.99), (2.101), (2.103), (2.104), and (2.105)), and its special cases, including viscoelastic solid, thermoelastic solid, elastic solid, and heat-conducting fluid. We have also demonstrated that Wang's representation theorem for isotropic functions is valuable in constructing constitutive equations for fluids and isotropic materials.

Remarks. The spirit of continuum mechanics is that it applies equally well to all kinds of continuous media, including gas, liquid, and solid, in both equilibrium and nonequilibrium systems.

Problems

1. Let $\mathbf{A}$ be an arbitrary second-order tensor, $\mathrm{tr}\,\mathbf{A} = A_{ii}$, $\mathrm{tr}\,\mathbf{A}^2 = A_{ij}A_{ji}$, $tr\,\mathbf{A}^3 = A_{ij}A_{jk}A_{ki}$.

   The Cayley–Hamilton theorem asserts that $\mathbf{A}$ satisfies its own characteristic equation:

$$\mathbf{A}^3 - I\mathbf{A}^2 + II\mathbf{A} - III\mathbf{I} = 0.$$

   Show that the three invariants can be expressed as

$$\begin{aligned} I &= \mathrm{tr}\,\mathbf{A}, \\ II &= \{(\mathrm{tr}\,\mathbf{A})^2 - \mathrm{tr}\,\mathbf{A}^2\}/2, \\ III &= \{(\mathrm{tr}\,\mathbf{A})^3 - 3\mathrm{tr}\,\mathbf{A}\mathrm{tr}\,\mathbf{A}^2 + 2\mathrm{tr}\,\mathbf{A}^3\}/6. \end{aligned}$$

2. Use the result
   det **A** det **B** = det ($\mathbf{A}^T\mathbf{B}$)
   to show that

$$\varepsilon_{ijk}\varepsilon_{lmn} = \delta_{il}(\delta_{jm}\delta_{kn} - \delta_{jn}\delta_{km}) + \delta_{im}(\delta_{jn}\delta_{kl} - \delta_{jl}\delta_{kn}) + \delta_{in}(\delta_{jl}\delta_{km} - \delta_{jm}\delta_{kl})$$

   and consequently

$$\varepsilon_{ijk}\varepsilon_{lmk} = \delta_{il}\delta_{jm} - \delta_{im}\delta_{jl},$$
$$\varepsilon_{ijk}\varepsilon_{ljk} = 2\delta_{il}.$$

3. Prove the following vector identities involving ∇:

$$\nabla\cdot(\nabla F) = \nabla^2 F,$$
$$\nabla(FG) = F\nabla G + G\nabla F,$$
$$\nabla^2(FG) = F\nabla^2 G + 2(\nabla F)\cdot(\nabla G) + G\nabla^2 F,$$
$$\nabla\cdot(F\mathbf{v}) = (\nabla F)\cdot\mathbf{v} + F\nabla\cdot\mathbf{v},$$
$$\nabla\cdot(F\nabla G) = F\nabla^2 G + \nabla F\cdot\nabla G,$$
$$\nabla\times(\nabla F) = 0,$$
$$\nabla\cdot(\nabla\times\mathbf{v}) = 0,$$
$$\nabla\cdot(\mathbf{a}\times\mathbf{b}) = (\nabla\times\mathbf{a})\cdot\mathbf{b} - \mathbf{a}\cdot(\nabla\times\mathbf{b}),$$
$$\nabla\times(F\mathbf{v}) = \nabla F\times\mathbf{v} + F\nabla\times\mathbf{v},$$
$$\nabla\times(\nabla\times\mathbf{v}) = \nabla(\nabla\cdot\mathbf{v}) - \nabla^2\mathbf{v}.$$

4. **A** is a second-order tensor, verify that

$$j \equiv \det(\mathbf{A}) = \varepsilon_{ijk}\varepsilon_{lmn}A_{il}A_{jm}A_{kn}/6,$$
$$A_{li}^{-1} = \frac{1}{2j}\varepsilon_{ijk}\varepsilon_{lmn}A_{jm}A_{kn},$$
$$\frac{\partial j}{\partial A_{ij}} = jA_{ji}^{-1}.$$

5. Verify that

$$j \equiv \det(x_{k,K}) = e_{klm}e_{KLM}x_{k,K}x_{l,L}x_{m,M}/6,$$
$$X_{K,k} = \frac{e_{klm}e_{KLM}x_{l,L}x_{m,M}}{2j}.$$

6. Prove that

$$\frac{\partial j}{\partial x_{k,K}} = jX_{K,k},$$
$$(jX_{K,k})_{,K} = 0,$$
$$(j^{-1}x_{k,K})_{,k} = 0.$$

7. The Cauchy deformation tensor and the Green deformation tensor are defined, respectively, as follows:

$$c_{kl} \equiv X_{K,k}X_{K,l},$$
$$C_{KL} \equiv x_{k,K}x_{k,L}.$$

The Eulerian strain tensor is defined as $\varepsilon_{kl} \equiv \frac{1}{2}(\delta_{kl} - c_{kl})$ and the Lagrangian strain tensor is defined as $E_{KL} \equiv \frac{1}{2}(C_{KL} - \delta_{KL})$. Show that

$$E_{KL} = \varepsilon_{kl}x_{k,K}x_{l,L},$$
$$\varepsilon_{kl} = E_{KL}X_{K,k}X_{L,l}.$$

8. From $\boldsymbol{u} = \boldsymbol{x} - \boldsymbol{X} + \boldsymbol{b}$, show that

$$U_K = \delta_{Kl}x_l - X_K + B_K,$$
$$u_k = x_k - \delta_{Lk}X_L + b_k,$$
$$\mathrm{d}x_k = (\delta_{MK} + U_{M,K})\delta_{Mk}\mathrm{d}X_K,$$
$$\mathrm{d}X_K = (\delta_{mk} - u_{m,k})\delta_{mK}\mathrm{d}x_k,$$
$$2E_{KL} = U_{K,L} + U_{L,K} + U_{M,K}U_{M,L},$$
$$2\varepsilon_{kl} = u_{k,l} + u_{l,k} - u_{m,k}u_{m,l}.$$

9. Prove that

$$\mathrm{d}v = j\mathrm{d}V,$$
$$\mathrm{d}a_k = jX_{K,k}\mathrm{d}A_K,$$
$$\mathrm{d}A_K = j^{-1}x_{k,K}\mathrm{d}a_k.$$

10. For C_{KL} to play the role of a metric tensor, the Riemann–Christoffel tensor formed from it must vanish, i.e.,

$$R^{(\mathrm{C})}_{KLMN} = 0.$$

Find the compatibility equations in terms of E_{KL}.

11. Prove that

$$\frac{\mathrm{d}}{\mathrm{d}t}(x_{k,K}) = v_{k,l}x_{l,K},$$
$$\frac{\mathrm{d}}{\mathrm{d}t}(X_{K,k}) = -v_{l,k}X_{K,l},$$
$$\frac{\mathrm{d}}{\mathrm{d}t}(\mathrm{d}s^2) = 2v_{k,l}\mathrm{d}x_k\mathrm{d}x_l = 2d_{kl}x_{k,K}x_{l,L}\mathrm{d}X_K\mathrm{d}X_L,$$
$$\dot{E}_{KL} = d_{kl}x_{k,K}x_{l,L},$$
$$\frac{\mathrm{d}j}{\mathrm{d}t} = jv_{k,k},$$
$$\frac{\mathrm{d}}{\mathrm{d}t}(\mathrm{d}a_k) = v_{m,m}\mathrm{d}a_k - v_{m,k}\mathrm{d}a_m.$$

12. Determine the objectivity of the following quantities:
(a) x_k (Eulerian coordinate)
(b) v_k (velocity)
(c) $a_k = \dot{v}_k$ (acceleration)
(d) $c_{kl} = X_{K,k}X_{K,l}$ (Cauchy's deformation tensor)
(e) $v_{k,l}$ (velocity gradient)
(f) $\omega_{kl} = (v_{k,l} - v_{l,k})/2$ (spin tensor)
(g) $d_{kl} = (v_{k,l} + v_{l,k})/2$ (deformation rate tensor)
(h) $x_{k,K}$ (deformation gradient)
(i) $C_{KL} = x_{k,K}x_{k,L}$ (Green's deformation tensor)
(j) $\mathrm{d}a_k$ (differential area vector)
(k) $\frac{\mathrm{d}}{\mathrm{d}t}(\mathrm{d}a_k)$ (material time rate of $\mathrm{d}a_k$)
(l) t_{kl} (Cauchy's stress tensor)
(m) $\dot{t}_{kl}$ (material time rate of Cauchy's stress tensor)
(n) $T_{Km} \equiv jX_{K,k}t_{km}$ (first-order Piola–Kirchhoff stress tensor)
(o) $T_{KL} \equiv jX_{K,k}X_{L,l}t_{kl}$ (second-order Piola–Kirchhoff stress tensor)
(p) $A_{kl}^{(M)}$ (Rivlin–Ericksen tensor of order M)
(q) $C_{KL}^{(M)}$ (Mth order material time derivative of Green's deformation tensor)

Note that Rivlin–Ericksen tensor of order 1 is defined as

$$A_{kl}^{(1)} \equiv d_{kl}$$

and the $M+1$ order Rivlin–Ericksen tensor is defined as ($M = 1, 2, 3, \ldots..$)

$$A_{kl}^{(M+1)} \equiv \frac{\mathrm{d}}{\mathrm{d}t}A_{kl}^{(M)} + A_{km}^{(M)}v_{m,l} + A_{ml}^{(M)}v_{m,k}.$$

and $\boldsymbol{A}^{(M)}$ and $\boldsymbol{C}^{(M)}$ are related as

$$C_{KL}^{(M)} = A_{kl}^{(M)}x_{k,K}x_{l,L}.$$

13. Tensor a_{ij} is objective,
(a) determine the objectivity of $\dot{a}_{ij}$
(b) the Jaumann rate of **a** is defined as

$$\hat{a}_{ij} \equiv \dot{a}_{ij} - \omega_{ik}a_{kj} + a_{ik}\omega_{kj},$$

is $\hat{\mathbf{a}}$ objective?
(c) the Truesdell rate of **a** is defined as

$$a_{ij}^{*} \equiv \dot{a}_{ij} - v_{i,k}a_{kj} - v_{j,k}a_{ik} + a_{ij}v_{k,k},$$

is $\mathbf{a}^*$ objective?

14. Verify that

$$T_{Km}\mathrm{d}A_K = t_{km}\mathrm{d}a_k.$$

What is the physical meaning of T_{Km}? Why it is named the engineering stress?

15. Show that the balance law of linear momentum can be written as

$$T_{Kk,K} + \rho^0(f_k - \dot{v}_k) = 0.$$

Is it true that $T_{Kk} = T_{kK}$?

16. Show that the balance law of linear momentum can also be expressed in terms of the second-order Piola–Kirchhoff stress tensor as

$$(T_{KL}x_{k,L})_{,K} + \rho^0(f_k - \dot{v}_k) = 0.$$

Is it true that $T_{KL} = T_{LK}$?

17. Find the detailed expressions for the following constitutive relations based on Wang's representation theorem for isotropic functions:
(a) $\psi = \psi(x_{k,K}, \theta_{,k}, X_K)$
(b) $q_m = q_m(A_{ij}, B_{ij}, v_i, \theta)$,
where **A** and **B** are first- and second-order Rivlin–Ericksen tensors, respectively, and
$v_i \equiv \frac{\partial\theta}{\partial x_i}$ is the temperature gradient.
(c) $T_{KL} = T_{KL}(x_{m,M})$
(d) $\psi = \psi(x_{m,M}, x_{m,MN})$
(e) $t_{ij} = t_{ij}(v_{k,l}, \theta_{,k})$
where $v_{k,l}$ is the velocity gradient. Note that it is neither symmetric nor antisymmetric.

18. Consider the material is isotropic. To begin with, the constitutive relations have the following forms. Based on Wang's representation theorem, further derive each of the following constitutive relations.
(a) $\psi = \psi(C_{KL}, \dot{C}_{KL}, \theta_{,K})$.
(b) $T_{KL} = T_{KL}(E_{MN}, \dot{E}_{MN}, \theta_{,M})$.
(c) $Q_K = Q_K(E_{MN}, \dot{E}_{MN}, \theta_{,M})$.

19. To begin with, a heat conducting fluid is assumed to have the following constitutive equations:

$$\begin{aligned}
\mathbf{t} &= \mathbf{t}(\rho^{-1}, \mathbf{d}, \theta, \nabla\theta),\\
\mathbf{q} &= \mathbf{q}(\rho^{-1}, \mathbf{d}, \theta, \nabla\theta),\\
\psi &= \psi(\rho^{-1}, \mathbf{d}, \theta, \nabla\theta),\\
\eta &= \eta(\rho^{-1}, \mathbf{d}, \theta, \nabla\theta).
\end{aligned}$$

Use CD inequality to prove that these constitutive equations are reduced to

$$\begin{aligned}
&\psi = \psi(\rho^{-1}, \theta),\\
&\eta = -\frac{\partial\psi}{\partial\theta},\\
&\pi \equiv -\frac{\partial\psi}{\partial\rho^{-1}},\\
&t_{kl} = -\pi\delta_{kl} + t^d_{kl}(\rho^{-1}, \theta, \mathbf{d}, \nabla\theta),\\
&t^d_{kl}d_{kl} + \frac{1}{\theta}q_k\theta_{,k} \geq 0.
\end{aligned}$$

Use Wang's representation theorem to show that

$$\begin{aligned}\mathbf{t}^d &= c_0\mathbf{I} + c_1\mathbf{d} + c_2\mathbf{d}^2 + c_3\mathbf{u}\otimes\mathbf{u} + c_4(\mathbf{u}\otimes\mathbf{du} + \mathbf{du}\otimes\mathbf{u}) \\ &\quad + c_5(\mathbf{u}\otimes\mathbf{d}^2\mathbf{u} + \mathbf{d}^2\mathbf{u}\otimes\mathbf{u}), \\ \mathbf{q}/\theta &= c_6\mathbf{u} + c_7\mathbf{du} + c_8\mathbf{d}^2\mathbf{u}.\end{aligned}$$

Find the constraints imposed by CD inequality.

Answer:

$$c_0 I_1 + c_1 I_2 + c_2 I_3 + c_3(I_4 + I_7) + c_4(I_5 + I_8) + c_5 I_6 + c_6 I_7 + c_7 I_8 \geq 0,$$

where

$$\begin{aligned} c_i &= c_i(\rho^{-1}, \theta, I_1, I_2, I_3, I_4, I_7, I_8) \quad (i = 0, 1, 2, \ldots, 7), \\ \mathbf{u} &\equiv \nabla\theta, \\ d_{ij}^2 &\equiv d_{ik}d_{kj}, \\ I_1 &\equiv d_{kk}, \\ I_2 &\equiv d_{kl}d_{lk}, \\ I_3 &\equiv d_{km}d_{ml}d_{lk}, \\ I_4 &\equiv \mathbf{u}\cdot\mathbf{du} = u_i d_{ij} u_j, \\ I_5 &\equiv (\mathbf{du})\cdot(\mathbf{du}) = d_{ij}u_j d_{ik}u_k, \\ I_6 &\equiv \mathbf{u}\cdot\mathbf{d}^3\mathbf{u} + (\mathbf{du})\cdot(\mathbf{d}^2\mathbf{u}), \\ I_7 &\equiv \mathbf{u}\cdot\mathbf{u}, \\ I_8 &\equiv \mathbf{u}\cdot\mathbf{d}^2\mathbf{u}. \end{aligned}$$

20. To begin with, a thermoviscoelastic solid is assumed to have the following constitutive equations:

$$\begin{aligned}\mathbf{T} &= \mathbf{T}(\mathbf{E}, \dot{\mathbf{E}}, \theta, \nabla\theta, \mathbf{X}), \\ \mathbf{Q} &= \mathbf{Q}(\mathbf{E}, \dot{\mathbf{E}}, \theta, \nabla\theta, \mathbf{X}), \\ \psi &= \psi(\mathbf{E}, \dot{\mathbf{E}}, \theta, \nabla\theta, \mathbf{X}), \\ \eta &= \eta(\mathbf{E}, \dot{\mathbf{E}}, \theta, \nabla\theta, \mathbf{X}).\end{aligned}$$

Find the constraints on these constitutive equations based on CD inequality.

21. Following Problem 20 and assuming that the material is isotropic, further derive the constitutive equations. Derive the energy equation.
22. Following Problem 21 and assuming that the constitutive relations between the independent and the dependent constitutive variables (except ψ) are linear, further derive the constitutive equations. Derive the energy equation.
23. Following Problem 20 and assuming that the constitutive relations between the independent and the dependent constitutive variables (except ψ) are linear, further derive the constitutive equations. Derive the energy equation.

24. Following Problem 23 and assuming that the material is isotropic, further derive the constitutive equations. Derive the energy equation.
25. What happens to those constitutive equations if the thermoviscoelastic solid is melted into fluid?
26. For isotropic elastic solid, the Cauchy stress tensor in the first approach is written as

$$t_{ij} = t_{ij}(x_{k,K}).$$

Further derive this constitutive equation based on the principle of objectivity and Wang's representation theorem. In the second approach, let

$$t_{kl} = \frac{2}{J}\frac{\partial \Sigma(\mathbf{C})}{\partial C_{KL}} x_{k,K} x_{l,L},$$

where **C** is the Green deformation tensor. Again, further derive this constitutive equation based on the principle of objectivity and Wang's representation theorem. Compare the results from these two approaches. Are there any similarity and difference? Why or why not?

3
Fundamentals of Finite Element Method

The fundamentals of continuum mechanics were discussed in Chapter 2. In this chapter, we first present the integral formulation of continuum problems, and then move on to present the essentials of finite element methods. The development is necessary as a prerequisite to the subsequent chapters in meshless methods.

Integral Formulation of Continuum Problems

Physical problem arising in engineering and science can often be represented by a set of differential equations and boundary conditions. The problem to be solved in most general cases is to find an unknown function **u** such that it satisfies a set of differential equations

$$\mathbf{D}(\mathbf{u}) = \begin{Bmatrix} D_1(\mathbf{u}) \\ D_2(\mathbf{u}) \\ D_3(\mathbf{u}) \\ \vdots \end{Bmatrix} = \mathbf{f} \tag{3.1}$$

in a domain Ω, together with a set of boundary conditions

$$\mathbf{B}(\mathbf{u}) = \begin{Bmatrix} B_1(\mathbf{u}) \\ B_2(\mathbf{u}) \\ B_3(\mathbf{u}) \\ \vdots \end{Bmatrix} = \mathbf{g} \tag{3.2}$$

on the boundaries Γ of the domain (Fig. 3.1). **D** and **B** are differential operators operating on the unknown function **u**. The boundary condition that does not involve differential operator is often named the essential boundary condition. The boundary condition that involves differential operator is referred to as nonessential or natural boundary condition.

The function **u** sought may be a scalar quantity or may represent a vector of several variables. Correspondingly, the differential equation may be a single equation or a set of simultaneous equations.

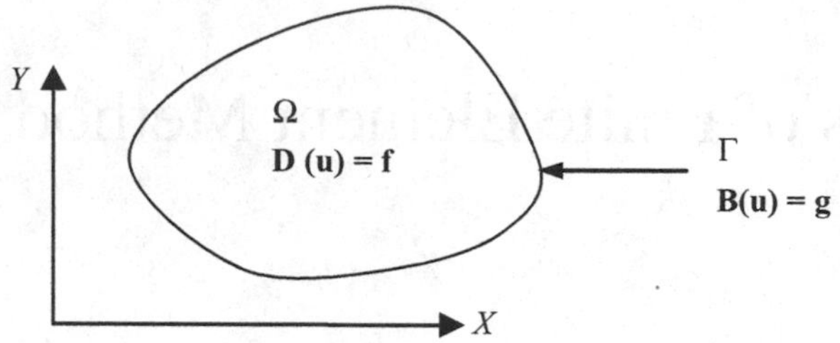

FIGURE 3.1. Problem domain Ω and its boundary Γ.

The problem for which a model is obtained using a finite number of well-defined components is termed *discrete*. If the model is continued indefinitely and the problem can only be defined using the mathematical fiction of an infinitesimal, and it leads to differential equations with an infinite number of elements. We term this *continuous*.

Continuous problem can only be solved exactly by mathematical manipulation. However, the available mathematical techniques usually limit the possibilities to oversimplified situations. With the advent of digital computers, discrete problems can generally be solved readily even if the number of elements is very large.

To overcome the intractability of the realistic type of continuum problem, various methods of *discretization* have been proposed and developed. All involve an *approximation* that is of such a kind that it approaches, as closely as desired, the true continuum solution as the number of discrete variables increases. We have had finite difference methods, finite volume methods, finite element methods, and boundary element methods, and now we have meshless methods.

Those above-mentioned methods seek the solution in the approximate form

$$\mathbf{u} \approx \hat{\mathbf{u}} = \Phi_i \mathbf{a}_i, \tag{3.3}$$

where the tensor summation convention is adopted; $i = 1, 2, \ldots n$; Φ_i may be called shape, basis, or interpolation functions prescribed in terms of independent variables, such as the coordinates x, y, and z; and all or some of $\mathbf{a}_i$ are unknown parameters. These shape or basis functions are usually defined locally for elements or subdomains. The properties of the discrete systems can be recovered if the approximating equations are cast in an integral form

$$\int_\Omega G_j(\mathbf{u})\, d\Omega + \int_\Gamma h_j(\mathbf{u})\, d\Gamma = 0 \quad j = 1, 2, 3, \ldots, n \tag{3.4}$$

in which G_j and h_j prescribe known functions or operators. These integral forms will permit the approximation to be obtained node by node and an assembly can be achieved as

$$\int_\Omega G_j(\mathbf{u})\, d\Omega + \int_\Gamma h_j(\mathbf{u})\, d\Gamma = \sum_i^m \left(\int_{\Omega_i} G_j\, d\Omega + \int_{\Gamma_i} h_j(\mathbf{u})\, d\Gamma \right), \tag{3.5}$$

where Ω_i is the domain of the ith element and Γ_i its part of the boundary.

Two distinct procedures are available for obtaining the approximation in such integral formulations: the weighted residual methods and the variational functionals.

A weighted residual method uses integral expressions that contain the differential equations of a physical problem. The Galerkin method, which is the most popular weighted residual method, is used to produce finite element and some meshless formulations.

A variational principle uses an integral expression, called a functional, that yields the governing differential equations and nonessential boundary conditions of a problem when operated upon by standard procedures of the calculus of variations. The principle of stationary potential energy is one of many variational principles.

Functional and residual formulations are both known as "weak" form of stating the governing equations of a problem. The differential equations themselves comprise the "strong" form. The weak form enforces conditions in an average or integral sense, whereas the strong form enforces them at every point.

Weighted Residual Methods

The governing differential equations and nonessential boundary conditions of an arbitrary physical problem are symbolized as

$$D(u) - f = 0 \text{ in domain } \Omega \tag{3.6}$$

$$B(u) - g = 0 \text{ on boundary } \Gamma \text{of } \Omega. \tag{3.7}$$

For example, a beam-bending problem, as shown in Fig. 3.2, can be described with

$$EIv_{,xxxx} = -q \quad x \in (0, L), \tag{3.8}$$

$$\left.\begin{aligned} EIv_{,xx} &= M_L \\ EIv_{,xxx} &= Q_L \end{aligned}\right\} \text{ at } x = L, \tag{3.9}$$

$$\left.\begin{aligned} v_{,x} &= 0 \\ v &= 0 \end{aligned}\right\} \text{ at } x = 0, \tag{3.10}$$

where v is the vertical displacement, q the distributed load, E Young's modulus, I the moment of inertia of the cross-sectional area, and EI the bending stiffness.

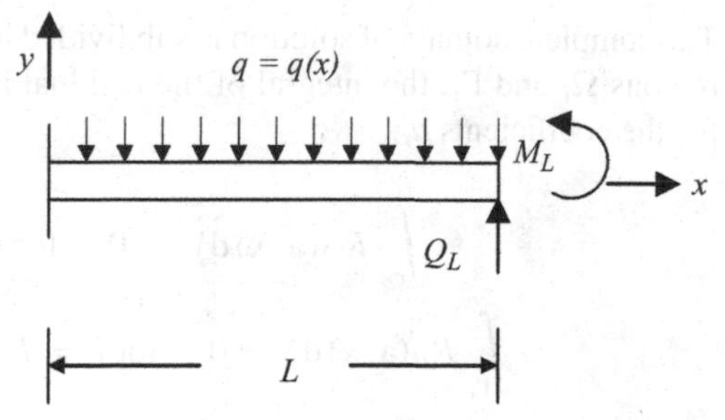

FIGURE 3.2. A beam-bending problem.

Equation (3.8) is the governing differential equation and can be symbolized by Eq. (3.6) with $D = EI\mathrm{d}^4/\mathrm{d}x^4$, $u = v$, and $f = -q$. The two equations in Eq. (3.9) are the nonessential boundary conditions, corresponding to Eq. (3.7) with $B = EI\mathrm{d}^2/\mathrm{d}x^2$, $g = M_L$ and $B = EI\mathrm{d}^3/\mathrm{d}x^3$, $g = Q_L$, respectively. The first equation in Eq. (3.10) is the nonessential boundary condition, corresponding to Eq. (3.7) with $B = \mathrm{d}/\mathrm{d}x$, $g = 0$ and note that the second equation in Eq. (3.10) is an essential boundary condition because the unknown function v itself is specified at the boundary.

In general, we seek an approximate solution, $\hat{u}$. Typically, $\hat{u}$ is a polynomial that satisfies essential boundary conditions and contains undermined coefficients $a_1, a_2, \ldots, a_n$. To obtain an approximate solution, we must determine the values of a_i ($i = 1, 2, 3, \ldots, n$) such that u and $\hat{u}$ are as close as possible.

If $\hat{u}$ is substituted into Eqs. (3.6) and (3.7), equality does not prevail because $\hat{u}$ is not exact. The discrepancy can be expressed as residuals R_D and R_B, which are functions of $\mathbf{x}$ and the a_i:

$$R_D = R_D(\mathbf{a}, \mathbf{x}) = D(\hat{u}) - f, \tag{3.11}$$

$$R_B = R_B(\mathbf{a}, \mathbf{x}) = B(\hat{u}) - g. \tag{3.12}$$

We presume that $\hat{u}$ is a good approximation of u if residuals are small. Small residuals can be achieved by various methods, each of which is designed to produce algebraic equations that can be solved for the n coefficients a_i. The various common choices are summarized as follows.

Point Collocation

For n distinct points $\mathbf{x}_i$ *in* the solution domain, the residuals are set to zero to obtain n simultaneous equations for the n coefficients a_i, i.e.,

$$R_D(\mathbf{a}, \mathbf{x}_i) = 0 \quad \text{for } i = 1, 2, 3, \ldots, j, \tag{3.13}$$

$$R_B(\mathbf{a}, \mathbf{x}_i) = 0 \quad \text{for } i = j+1, j+2, \ldots, n. \tag{3.14}$$

Subdomain Collocation

The complete domain of solution is subdivided into n subdomains. Over n different regions Ω_i and Γ_i, the integral of the residual is set to zero to obtain n equations for the coefficients a_i.

$$\int_{\Omega_i} R_D(\mathbf{a}, \mathbf{x})\,\mathrm{d}V = 0 \quad \text{for } i = 1, 2, 3, \ldots, j, \tag{3.15}$$

$$\int_{\Gamma_i} R_B(\mathbf{a}, \mathbf{x})\,\mathrm{d}S = 0 \quad \text{for } i = J+1, J+2, \ldots, n. \tag{3.16}$$

Continuous Least Squares

The a_is are chosen to minimize a function I:

$$\frac{\partial I}{\partial a_i} = 0 \quad \text{for } i = 1, 2, 3, \ldots n. \tag{3.17}$$

Function I is formed by integrating squares of the residuals,

$$I = \int_{\Omega} [R_D(\mathbf{a}, \mathbf{x})]^2 \, \mathrm{d}V + \alpha \int_{\Gamma} [R_B(\mathbf{a}, \mathbf{x})]^2 \, \mathrm{d}S \tag{3.18}$$

where α is an arbitrary scalar multiplier that may be used to achieve dimensional homogeneity and also serve as a penalty number. Large values of α increase the importance of R_B relative to R_D.

Least Squares Collocation

The $a_i (i = 1, 2, 3, \ldots, n)$ is still chosen to minimize a function I, but I is defined in terms of squared residuals at m points $x_j, (j = 1, 2, 3, \ldots, m)$. The number of points is larger than the number of coefficients $a_i, (i = 1, 2, 3, \ldots, n)$, i.e., $m \geq n$. We have

$$I = \sum_{j=1}^{k-1} [R_D(\mathbf{a}, \mathbf{x}_j)]^2 + \alpha \sum_{j=k}^{m} [R_B(\mathbf{a}, \mathbf{x}_j)]^2. \tag{3.19}$$

$$\frac{\partial I}{\partial a_i} = 0 \quad \text{for } i = 1, 2, 3, \ldots n. \tag{3.20}$$

Equation (3.20) yields n equations for a_i, even for $m > n$. The method is also called *point least squares* or *overdetermined collocation.* If $m = n$, the method becomes simple point collocation.

Galerkin

In this technique, the coefficients a_i are determined from the n equations of weighted residuals

$$R_i = \int_{\Omega} W_i(\mathbf{x}) R_D(\mathbf{a}, \mathbf{x}) \, \mathrm{d}V = 0 \ \text{for } i = 1, 2, 3, \ldots n. \tag{3.21}$$

In the general Galerkin method, also called Bubnov–Galerkin method, the weight functions, $W_i = W_i(\mathbf{x})$, are coefficients of the generalized coordinates a_i. Thus $W_i = \partial \hat{u} / \partial a_i$. In the Petrov–Galerkin method, other forms of W_i are used. Galerkin method leads frequently to symmetric matrices. This is one reason that it has been adopted in finite element method almost exclusively. It is also widely used in meshless methods.

The commonality shared by the foregoing methods is that they all can be symbolized as

$$\int_{\Omega} W_i R \, \mathrm{d}\Omega = 0, \tag{3.22}$$

where R represents R_B and/or R_D, Ω represents Ω and/or Γ. It says that over the region of interest, the weighted residual has an average value of zero. The various weighted residuals differ in how W_i is defined. In the collocation and subdomain methods, the W_i is unit delta or step function that is nonzero at certain points or over certain regions. In least squares methods, $W_i = \partial R/\partial a_i$. In the Galerkin method, $W_i = \partial \hat{u}/\partial a_i$.

Example 1. Formulation of elasticity by Galerkin method.

In general three-dimensional continuum, the equilibrium equations can be written in tensor notation as

$$t_{ji,j} + \rho f_i - \rho \dot{v}_i = 0, \tag{3.23}$$

where $\mathbf{t}$ is the stress, $\rho\mathbf{f}$ the body force, and $\mathbf{v}$ the velocity. To obtain a weak form, we shall introduce a weighting tensor function $\Phi_{i\alpha}$. The Galerkin method then gives

$$\int_{\Omega} \Phi_{i\alpha}(t_{ji,j} + \rho f_i - \rho \dot{v}_i) \, \mathrm{d}V = 0. \tag{3.24}$$

Using the Green–Gauss theorem and integration by parts, the above equation becomes

$$\int_{\Omega} (t_{ji}\Phi_{i\alpha,j} + \rho \dot{v}_i \Phi_{i\alpha} - \rho f_i \Phi_{i\alpha}) \, \mathrm{d}V - \oint_{\Gamma} t_{ji} n_j \Phi_{i\alpha} \, \mathrm{d}S = 0. \tag{3.25}$$

For small strain elasticity,

$$e_{ij} = (u_{i,j} + u_{j,i})/2, \tag{3.26}$$

$$t_{ij} = t_{ji} = a_{ijkl} e_{kl}, \tag{3.27}$$

$$\dot{v}_i = \ddot{u}_i. \tag{3.28}$$

With the approximation, $\hat{u}_i = \Phi_{i\alpha} U_\alpha$, we have

$$\hat{e}_{ij} = \frac{1}{2}(\Phi_{i\alpha,j} + \Phi_{j\alpha,i}) U_\alpha \triangleq B_{ij\alpha} U_\alpha. \tag{3.29}$$

Equation (3.25) then becomes

$$\int_{\Omega} (a_{ijkl} B_{kl\beta} U_\beta \Phi_{i\alpha,j} + \rho \Phi_{i\beta} \ddot{U}_\beta \Phi_{i\alpha} - \rho f_i \Phi_{i\alpha}) \, \mathrm{d}V - \oint_{\Gamma} t_{ji} n_j \Phi_{i\alpha} \, \mathrm{d}S = 0, \tag{3.30}$$

which can be rewritten as

$$U_\beta \int_{\Omega} a_{ijkl} B_{kl\beta} B_{ij\alpha} \, \mathrm{d}V + \ddot{U}_\beta \int_{\Omega} \rho \Phi_{i\beta} \Phi_{i\alpha} \, \mathrm{d}V - \int_{\Omega} \rho f_i \Phi_{i\alpha} \, \mathrm{d}V$$
$$- \int_{\Gamma_t} \bar{t}_i \Phi_{i\alpha} \, \mathrm{d}S - \int_{\Gamma_u} t_{ji} n_j \Phi_{i\alpha} \, \mathrm{d}S = 0. \tag{3.31}$$

The fourth term is the natural boundary condition, i.e., $n_j t_{ji} = \bar{t}_i$. The fifth term is the essential boundary condition and it vanishes in the methods where approximation functions pass through nodes, such as finite element method. Equation (3.31) results a finite element formulation

$$M_{\alpha\beta}\ddot{U}_\beta + K_{\alpha\beta}U_\beta = F_\alpha \quad \text{or} \quad \mathbf{M}\ddot{\mathbf{U}} + \mathbf{K}\mathbf{U} = \mathbf{F}, \tag{3.32}$$

where

$$\begin{aligned} M_{\alpha\beta} &= \int_\Omega \rho \Phi_{i\beta}\Phi_{i\alpha}\,\mathrm{d}V, \\ K_{\alpha\beta} &= \int_\Omega a_{ijkl}B_{kl\beta}B_{ij\alpha}\,\mathrm{d}V, \\ F_\alpha &= \int_\Omega \rho f_i \Phi_{i\alpha}\,\mathrm{d}V + \int_{\Gamma_t} \bar{t}_i \Phi_{i\alpha}\,\mathrm{d}S. \end{aligned} \tag{3.33}$$

In meshless methods, the fifth term in Eq. (3.31) remains, additional treatment of essential boundary condition is needed, and this will be discussed in next section.

If we replace $\Phi_{i\alpha}$ by δu_i, then Eq. (3.25) becomes

$$\int_\Omega (t_{ji}\delta u_{i,j} + \rho \dot{v}_i \delta u_i)\,\mathrm{d}V - \oint_\Gamma t_{ji} n_j \delta u_i\,\mathrm{d}S = 0. \tag{3.34}$$

This is the virtual work statement. Therefore, the method of virtual work is in fact a Galerkin formulation of the weighted residual process applied to the equilibrium equation.

Variational Principle

A variational principle specifies a scalar function Π that is defined by an integral form

$$\Pi = \int_\Omega F(u_i, u_{i,j}, \ldots)\,\mathrm{d}V + \int_\Gamma E(u_i, u_{i,j}, \ldots)\,\mathrm{d}S, \tag{3.35}$$

in which $\mathbf{u}$ is the unknown function and F and E are the specified differential operators. The solution to the continuum problem is a function $\mathbf{u}$ that makes Π stationary with respect to small changes $\delta\mathbf{u}$. Thus, for a solution to the continuum problem, the "variation" is

$$\delta\Pi = 0. \tag{3.36}$$

Assuming a trial function

$$u_i \approx \hat{u}_i = \Phi_{i\alpha} a_\alpha, \tag{3.37}$$

it follows that

$$\delta\Pi = \frac{\partial \Pi}{\partial a_i}\delta a_i. \tag{3.38}$$

Since the above equation should hold for any arbitrary variation δa_i, this implies, for each a_i,

$$\frac{\partial \Pi}{\partial a_i} = 0. \tag{3.39}$$

We can then have a set of equations for a_i. If a "variational principle" can be found, the approximate solutions in the standard integral form can then be obtained. In the case that the functional Π is quadratic function of a_i, then Eq. (3.39) yields a set of linear algebra equations

$$\frac{\partial \Pi}{\partial \mathbf{a}} = \mathbf{K}\mathbf{a} - \mathbf{f}. \tag{3.40}$$

Zienkiewicz (1983) has shown that the matrix $\mathbf{K}$ will always be symmetric. In fact, the symmetry matrices will arise whenever a variational principle exists in one of the most important merits of variational approaches for discretization. It is noted that symmetric matrices can also frequently arise directly from the Galerkin process.

Some physical problems can be stated directly in a variational principle form. There are some others for which the variational principle has to be constructed by considering some constraints. We called them the constrained variational principles. Two common choices are the Lagrange multiplier method and the penalty functions method.

Lagrange Multipliers

Consider the problem of making a functional Π stationary, subject to that the unknown $\mathbf{u}$ obeying some set of additional differential relationships

$$\mathbf{C}(\mathbf{u}) = 0 \quad \text{in}\,\Omega, \tag{3.41}$$

$$\mathbf{E}(\mathbf{u}) = 0 \quad \text{on}\,\Gamma. \tag{3.42}$$

We can introduce such constraints by forming another functional

$$\bar{\Pi}(\mathbf{u}, \boldsymbol{\lambda}, \gamma) = \Pi(\mathbf{u}) + \int_\Omega \boldsymbol{\lambda} \cdot \mathbf{C}(\mathbf{u})\,\mathrm{d}V + \int_\Gamma \gamma \cdot \mathbf{E}(\mathbf{u})\,\mathrm{d}S, \tag{3.43}$$

in which $\boldsymbol{\lambda}$ and γ are functions of the coordinates and called as Lagrange multipliers. Note that $\boldsymbol{\lambda}$ and γ are vectors having same number of elements as in $\mathbf{C}(\mathbf{u})$ and $\mathbf{E}(\mathbf{u})$, respectively. The variation of the new functional is now

$$\delta\bar{\Pi} = \delta\Pi + \int_\Omega \delta\boldsymbol{\lambda} \cdot \mathbf{C}(\mathbf{u})\mathrm{d}V + \int_\Omega \boldsymbol{\lambda} \cdot \delta\mathbf{C}(\mathbf{u})\,\mathrm{d}V + \int_\Gamma \delta\gamma \cdot \mathbf{E}(\mathbf{u})\,\mathrm{d}S + \int_\Gamma \gamma \cdot \delta\mathbf{E}(\mathbf{u})\,\mathrm{d}S, \tag{3.44}$$

and this is zero providing $\mathbf{C}(\mathbf{u}) = 0$, $\mathbf{E}(\mathbf{u}) = 0$, and hence $\delta\mathbf{C}(\mathbf{u}) = 0$, $\delta\mathbf{E}(\mathbf{u}) = 0$, and simultaneously,

$$\delta\Pi = 0. \tag{3.45}$$

Penalty Functions

For penalty methods, the new functional is constructed as

$$\bar{\Pi}(\mathbf{u}, \alpha, \beta) = \Pi(\mathbf{u}) + \alpha \int_\Omega \mathbf{C}(\mathbf{u}) \cdot \mathbf{C}(\mathbf{u})\, dV + \beta \int_\Gamma \mathbf{E}(\mathbf{u}) \cdot \mathbf{E}(\mathbf{u})\, dS, \quad (3.46)$$

in which α and β are penalty numbers. If Π is a minimum of the solution, then α and β should be positive numbers. The solution obtained by the stationarity of the function $\bar{\Pi}$ will satisfy the constraints only approximately. The larger the values of α and β, the better will be the constraints achieved.

It is noted that introducing Lagrange multiplier allows constrained variational principles to be obtained at the expense of increasing the total number of unknowns. Also, even in linear problems the algebraic equations that have to be solved are now complicated by having zero diagonal terms (Zienkiewicz and Taylor, 1989). The penalty function method does not have these drawbacks, but the solution is often not as accurate as that by Lagrange multiplier method.

Example 2. Formulation of elasticity by variational principle.

Consider again the previous example of elasticity with the governing equation

$$t_{ji,j} + \rho f_i - \rho \dot{v}_i = 0, \quad (3.47)$$

and essential and natural boundary conditions

$$u_i = \bar{u}_i \quad \text{on } \Gamma_u, \quad (3.48)$$

$$t_i = \bar{t}_i \quad \text{on } \Gamma_t. \quad (3.49)$$

The strain energy and kinetic energy in a body of volume Ω for a linear elasticity are given, respectively, by

$$I = \frac{1}{2} \int_\Omega t_{ij} e_{ij}\, dV, \quad (3.50)$$

$$K = \frac{1}{2} \int_\Omega \rho v_i v_i\, dV. \quad (3.51)$$

The conservation of energy gives the potential energy as

$$\Pi \equiv \frac{1}{2} \int_\Omega t_{ij} e_{ij}\, dV + \frac{1}{2} \int_\Omega \rho v_i v_i\, dV - \oint_\Gamma t_i u_i\, dS - \int_\Omega \rho f_i u_i\, dV, \quad (3.52)$$

in which the third term is the work done by external force and the fourth term is the work done by body force. To enforce the essential boundary condition, we construct a new functional with Lagrange multiplier $\boldsymbol{\lambda}$

$$\bar{\Pi} \equiv \frac{1}{2} \int_\Omega t_{ij} e_{ij}\, dV + \frac{1}{2} \int_\Omega \rho v_i v_i\, dV - \int_{\Gamma_t} \bar{t}_i u_i\, dS - \int_\Omega \rho f_i u_i\, dV + \int_{\Gamma_u} \lambda_i (u_i - \bar{u}_i)\, dS. \quad (3.53)$$

Since

$$t_{ij} = \frac{\partial I}{\partial e_{ij}}, \tag{3.54}$$

$$v_i \delta v_i = \dot{u}_i \delta \dot{u}_i = \dot{u}_i \ddot{u}_i \delta t = \ddot{u}_i \delta u_i, \tag{3.55}$$

we have

$$\begin{aligned}
\delta \bar{\Pi} \equiv & \int_\Omega t_{ij} \delta e_{ij} \, \mathrm{d}V + \int_\Omega \rho v_i \delta v_i \, \mathrm{d}V - \int_{\Gamma_t} \bar{t} \delta u_i \, \mathrm{d}S_i \\
& - \int_\Omega \rho f_i \delta u_i \, \mathrm{d}V + \int_{\Gamma_u} \lambda_i \delta (u_i - \bar{u}_i) \, \mathrm{d}S + \int_{\Gamma_u} \delta \lambda_i (u_i - \bar{u}_i) \, \mathrm{d}S \\
= & \int_\Omega t_{ij} \delta e_{ij} \, \mathrm{d}V + \int_\Omega \rho \ddot{u}_i \delta u_i \, \mathrm{d}V - \int_{\Gamma_t} \bar{t}_i \delta u_i \, \mathrm{d}S \\
& - \int_\Omega \rho f_i \delta u_i \, \mathrm{d}V + \int_{\Gamma_u} \lambda_i \delta u_i \, \mathrm{d}S + \int_{\Gamma_u} \delta \lambda_i (u_i - \bar{u}_i) \, \mathrm{d}S.
\end{aligned} \tag{3.56}$$

Now approximate $\boldsymbol{\lambda}$ on Γ_u in terms of nodal value $\boldsymbol{\Lambda}$ as

$$\lambda_j = \psi_{j\alpha} \Lambda_\alpha, \quad \alpha = 1, 2, 3, \ldots, l, \tag{3.57}$$

where l is the number of nodes whose weight functions are nonzero on the essential boundary. Together with the constitutive relations, $t_{ij} = t_{ji} = a_{ijkl} e_{kl}$, and the approximation $\hat{u}_i = \Phi_{i\alpha} U_\alpha$, $\hat{e}_{ij} = \frac{1}{2}(\Phi_{i\alpha,j} + \Phi_{j\alpha,i}) U_\alpha = B_{ij\alpha} U_\alpha$, Eq. (3.56) becomes

$$\begin{aligned}
\delta \bar{\Pi} = & \int_\Omega (a_{ijkl} B_{kl\beta} U_\beta B_{ij\alpha} \delta U_\alpha + \rho \Phi_{i\beta} \ddot{U}_\beta \Phi_{i\alpha} \delta U_\alpha) \, \mathrm{d}V \\
& - \int_{\Gamma_t} \bar{t}_i \Phi_{i\alpha} \delta U_\alpha \, \mathrm{d}S - \int_\Omega \rho f_i \Phi_{i\alpha} \delta U_\alpha \, \mathrm{d}V \\
& + \int_{\Gamma_u} \psi_{i\beta} \Lambda_\beta \Phi_{i\alpha} \delta U_\alpha \, \mathrm{d}S + \int_{\Gamma_u} \psi_{i\beta} \delta \Lambda_\beta (\Phi_{i\alpha} U_\alpha - \bar{u}_i) \, \mathrm{d}S \\
= & \, 0.
\end{aligned} \tag{3.58}$$

For finite element methods,

$$\left.\begin{aligned} \delta U_\alpha &= 0 \\ \Phi_{i\alpha} U_\alpha - \bar{u}_i &= 0 \end{aligned}\right\} \quad \text{on } \Gamma_u; \tag{3.59}$$

hence $\partial \bar{\Pi} / \partial a_i = 0$ leads to

$$U_\beta \int_\Omega a_{ijkl} B_{kl\beta} B_{ij\alpha} \, \mathrm{d}V + \ddot{U}_\beta \int_\Omega \rho \Phi_{i\beta} \Phi_{i\alpha} \, \mathrm{d}V - \int_{\Gamma_t} \bar{t}_i \Phi_{i\alpha} \, \mathrm{d}S - \int_\Omega \rho f_i \Phi_{i\alpha} \, \mathrm{d}V = 0, \tag{3.60}$$

which is identical with Eq. (3.32) obtained previously.

For meshless methods, Eq. (3.59) does not hold, we have instead

$$\begin{aligned}U_\beta \int_\Omega a_{ijkl} B_{kl\beta} B_{ij\alpha}\, \mathrm{d}V + \ddot{U}_\beta \int_\Omega \rho \Phi_{i\beta}\Phi_{i\alpha}\, \mathrm{d}V + \Lambda_\beta \int_{\Gamma_u} \psi_{i\beta}\Phi_{i\alpha}\, \mathrm{d}S \\ = \int_{\Gamma_t} \bar{t}_i \Phi_{i\alpha}\, \mathrm{d}S + \int_\Omega \rho f_i \Phi_{i\alpha}\, \mathrm{d}V,\end{aligned} \tag{3.61}$$

and

$$U_\alpha \int_{\Gamma_u} \psi_{i\beta}\Phi_{i\alpha}\, \mathrm{d}S = \int_{\Gamma_u} \psi_{i\beta}\bar{u}_i\, \mathrm{d}S. \tag{3.62}$$

Or in tensor form

$$\begin{aligned}\mathbf{M}\ddot{\mathbf{U}} + \mathbf{K}\mathbf{U} + \mathbf{G}\boldsymbol{\Lambda} = \mathbf{F}, \\ \mathbf{G}^t\mathbf{u} = \mathbf{f},\end{aligned} \tag{3.63}$$

where $\mathbf{G}^t$ stands for transpose of $\mathbf{G}$ and

$$\begin{aligned}G_{\alpha\beta} &= \int_{\Gamma_u} \Phi_{i\alpha}\psi_{i\beta}\, \mathrm{d}S, \\ f_\alpha &= \int_{\Gamma_u} \bar{u}_i \psi_{i\alpha}\, \mathrm{d}S.\end{aligned} \tag{3.64}$$

Example 3. Type of governing equations of linear elasticity.

For linear elasticity, the stress–strain relation can be expressed as

$$t_{ij} = a_{ijmn} u_{m,n}, \tag{3.65}$$

the governing equation, $t_{ji,j} + \rho f_i - \rho \dot{v}_i = 0$, then becomes

$$a_{jimn} u_{m,nj} + \rho f_i - \rho \ddot{u}_i = 0. \tag{3.66}$$

Since a_{ijmn} is positive definite and ρ is positive, we have $B^2 - 4AC = 4a_{ijmn}\rho > 0$ (cf. Appendix C). The governing equation of elastodynamics is hyperbolic and Eq. (3.66) represents wave equations.

For isotropic material in static case, Eq. (3.66) reduces to

$$\lambda u_{m,mi} + \mu(u_{i,jj} + u_{j,ij}) + \rho f_i = 0, \tag{3.67}$$

where λ and μ are Lamé constants. In the case $u_1 = u(x, y)$, $u_2 = u_3 = 0$, we have

$$\begin{aligned}&\lambda u_{m,m1} + \mu(u_{,jj} + u_{j,1j}) + \rho f_1 \\ &= \lambda u_{,11} + \mu(u_{,11} + u_{,22} + u_{,11}) + \rho f_1 \\ &= (\lambda + 2\mu) u_{1,11} + \mu u_{1,22} + \rho f_1 \\ &= 0.\end{aligned} \tag{3.68}$$

It is seen that $B^2 - 4AC = -4(\lambda + 2\mu)\mu < 0$ and the equation of the static elastic case is then elliptic.

Introduction to Finite Element Method

The development of finite element methods for the solution of practical engineering problems began with the advent of the digital computer. The essence of a finite element solution of an engineering problem is that a set of governing algebraic equation is established and solved. Hence, it was only through the use of the digital computer that this process could be rendered effective and given general applicability.

The name "finite element methods" was first used by Clough in 1960. It gained wide respectability and attention when it was recognized as having a sound mathematical foundation: it can be regarded as the solution of a variational problem by minimization of a functional. Thus, the method was seen as applicable to all field problems that can be cast in a variational form. Large general purpose finite element computer programs emerged during the late 1960s and early 1970s. Examples include NASTRAN and ANSYS that have been popular for almost half century.

The process of approximating the behavior of a continuum by "finite elements" can be described as follows:

1. The continuum is divided into a number of "finite elements."
2. The elements are assumed to be connected at nodal points situated on their boundaries. The displacements of these nodal points are the basic unknown parameters of the problem.
3. A set of functions is chosen to define uniquely the state of displacement within each "finite element" in terms of the nodal displacements.
4. The displacement functions then define uniquely the state of strain within an element also in terms of the nodal displacements. These strains, together with any initial strains and the constitutive properties of the material, define the state of stress throughout the element and, hence, also on its boundary.
5. A system of "forces" concentrated at the nodes and equilibrating the boundary stresses and any distributed loads is determined, resulting in a set of algebraic equations.
6. Solving the algebraic equations gives the displacements. Other information derivable from the displacement field, such as the strain tensor and the stress tensor through the constitutive equations, can also be obtained.

The approach outlined in the procedure is known as the displacement formulation.

Shape Functions in Finite Element Method

In the derivation of finite element models, the element displacements, $\mathbf{u}(x, y, z)$, are assumed in the form of polynomials in local element coordinates, x, y, and z, with undetermined constant coefficients. These displacements are found to be linear combinations of the element nodal point displacements. The essence of the finite element method is thus to achieve the approximated relationship between the element displacement at any point and the element nodal point displacements directly through the use *of shape function*, i.e., approximating the displacement **u** at any point within the element as

$$u_i \approx \hat{u}_i = N_\alpha U_{i\alpha} \quad (i = 1, 2, 3, \quad \alpha = 1, 2, 3, \ldots, m), \tag{3.69}$$

where m is the number of nodes per element; $U_{i\alpha}$ the ith component of the αth nodal displacements for a particular element. Equation (3.69) is sometimes written as

$$u_i \approx \hat{u}_i = N_{i\beta} U_\beta, \tag{3.70}$$

where $N_{i\beta}$ is the (i, β)th component of the $3 \times 3m$ matrix made of m shape functions $N_\alpha (\alpha = 1, 2, 3, \ldots, m)$ and $U_\beta (\beta = 1, 2, 3, \ldots, 3m)$ is the βth component of the vector of nodal displacements of the finite element.

A rigid translation of an element means

$$u_i = U_{i\alpha} = A_i, \tag{3.71}$$

where $\mathbf{A} = \{A_1, A_2, A_3\}$ is a constant vector of translation. Then, from Eq. (3.69), it follows that

$$\sum_{\alpha=1}^{n} N_\alpha = 1, \tag{3.72}$$

which is the character of shape functions, often referred to as *partition of unity*. The shape function defined in this way is named as *standard shape functions*. It depends on the number of nodes. If the approximation is expressed as a series in which the shape function N_α does not depend on the number of nodes in mesh, the shape function is then called *hierarchic shape functions*. The standard shape functions are the basis of most finite element programs. Here, we will only introduce the standard shape functions for two-dimensional rectangular elements and three-dimensional prism elements of linear, quadratic, and cubic types. For shape functions of other types of elements, interested readers are referred to the book by Zienkiewicz and Taylor (1989).

Two-Dimensional Rectangular Elements

Figure 3.3 shows the most frequently used rectangular elements. For the first element with four nodes, it is obvious that a product of the form

$$\frac{1}{4}(\xi + 1)(\eta + 1)$$

gives unity at top right corners where $\xi = \eta = 1$ and zero at all the other corners. Also, a linear variation of the shape function of all sides exists and hence continuity is satisfied (cf. Problems). Introducing new variables

$$\xi_0 = \xi \xi_i, \quad \eta_0 = \eta \eta_i, \quad i = 1, 2, 3, 4, \tag{3.73}$$

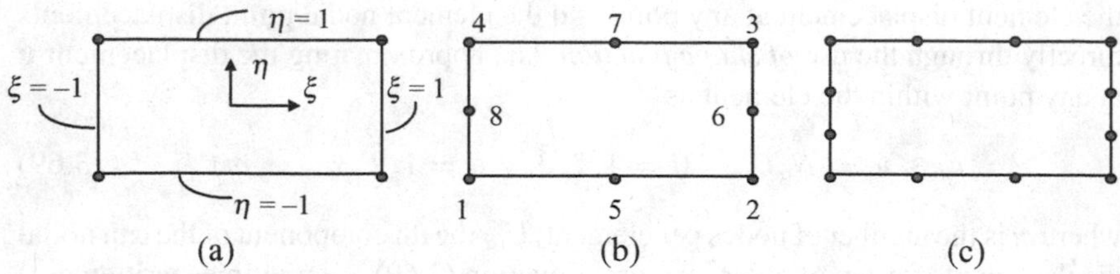

FIGURE 3.3. Most frequently used rectangular elements: (a) linear, (b) quadratic, (c) cubic.

where (ξ_i, η_i) are the coordinates of the ith node. Then the form

$$N_i = \frac{1}{4}(1+\xi_0)(1+\eta_0) \tag{3.74}$$

allows all shape functions to be written down in one expression.

It can be verified that

- For the 8-node element,

$$\text{for corner nodes: } N_i = \frac{1}{4}(1+\xi_0)(1+\eta_0)(\xi_0+\eta_0-1), \tag{3.75}$$

$$\text{for mid-side nodes: } \xi_i = 0 \quad N_i = \frac{1}{2}(1-\xi^2)(1+\eta_0), \tag{3.76}$$

$$\eta_i = 0 \quad N_i = \frac{1}{2}(1-\eta^2)(1+\xi_0). \tag{3.77}$$

- For the 12-node element,

$$\text{for corner nodes: } N_i = \frac{1}{32}(1+\xi_0)(1+\eta_0)(-10+9(\xi^2+\eta^2)), \tag{3.78}$$

$$\text{for mid-side nodes: } \xi_i = \pm 1 \text{ and } \eta_i = \pm\frac{1}{3} \tag{3.79}$$

$$N_i = \frac{9}{32}(1+\xi_0)(1-\eta^2)(1+9\eta_0),$$

$$\eta_i = \pm 1 \text{ and } \xi_i = \pm\frac{1}{3} \quad N_i = \frac{9}{32}(1+\eta_0)(1-\xi^2)(1+9\xi_0). \tag{3.80}$$

Three-Dimensional Solid Elements

For the most frequently used rectangular prism elements (Fig. 3.4), we have the following shape functions:

- 8-node linear element

$$N_i = \frac{1}{8}(1+\xi_0)(1+\eta_0)(1+\varsigma_0), \tag{3.81}$$

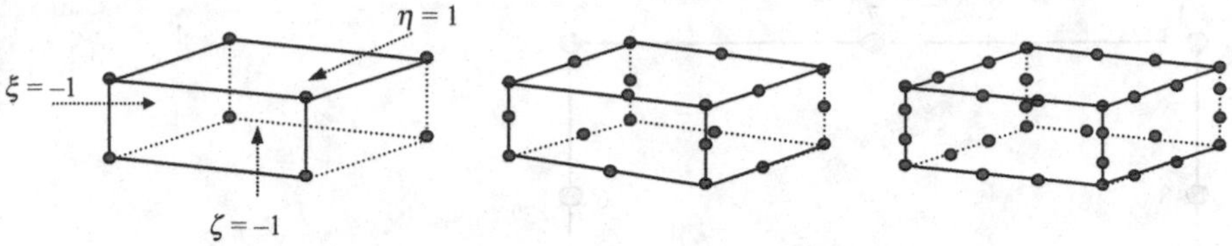

FIGURE 3.4. Most frequently used rectangular prism elements: (a) linear, (b) quadratic, (c) cubic.

- 20-node quadratic element

$$\text{Corner nodes: } N_i = \frac{1}{8}(1+\xi_0)(1+\eta_0)(1+\varsigma_0)(\xi_0+\eta_0+\varsigma_0-2), \tag{3.82}$$

Typical mid-side nodes: $\xi_i = 0 \quad \eta_i = \pm 1 \quad \varsigma_i = \pm 1$

$$N_i = \frac{1}{2}(1-\xi^2)(1+\eta_0)(1+\varsigma_0). \tag{3.83}$$

- 32-node cubic element

$$\text{Corner nodes: } N_i = \frac{1}{64}(1+\xi_0)(1+\eta_0)(1+\varsigma_0)[9(\xi^2+\eta^2+\varsigma^2)-19], \tag{3.84}$$

Typical mid-side nodes: $\xi_i = \pm\frac{1}{3} \quad \eta_i = \pm 1 \quad \varsigma_i = \pm 1$

$$N_i = \frac{9}{64}(1+\eta_0)(1+\varsigma_0)(1-\xi^2)(1+9\xi_0). \tag{3.85}$$

Example 4. Strain–displacement relation in two-dimensional finite element formulation.

For eight-node rectangular element, cf. Fig. 3.5 for nodal points, the approximated displacements in one element are

$$\begin{vmatrix} u_1 \\ u_2 \end{vmatrix} = \begin{vmatrix} N_1 & 0 & N_2 & 0 & N_3 & 0 & N_4 & 0 & N_5 & 0 & N_6 & 0 & N_7 & 0 & N_8 & 0 \\ 0 & N_1 & 0 & N_2 & 0 & N_3 & 0 & N_4 & 0 & N_5 & 0 & N_6 & 0 & N_7 & 0 & N_8 \end{vmatrix} \begin{vmatrix} U_{1x} \\ U_{1y} \\ U_{2x} \\ U_{-2y} \\ \vdots \\ U_{8x} \\ U_{8y} \end{vmatrix}, \tag{3.86}$$

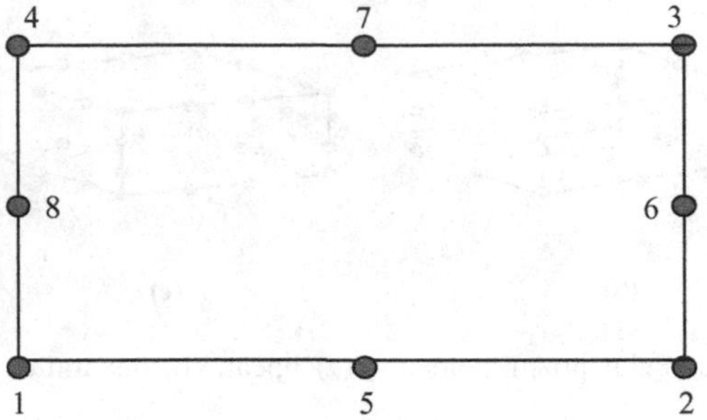

FIGURE 3.5. Two-dimensional element and its nodal points.

with

$$
\begin{aligned}
N_1 &= \frac{1}{4}(1-\xi)(1-\eta)(-\xi-\eta-1)\\
N_2 &= \frac{1}{4}(1+\xi)(1-\eta)(\xi-\eta-1)\\
N_3 &= \frac{1}{4}(1+\xi)(1+\eta)(\xi+\eta-1)\\
N_4 &= \frac{1}{4}(1-\xi)(1+\eta)(-\xi+\eta-1)\\
N_5 &= \frac{1}{2}(1-\xi)(1-\eta)(1+\xi)\\
N_6 &= \frac{1}{2}(1+\xi)(1-\eta)(1+\eta)\\
N_7 &= \frac{1}{2}(1-\xi)(1+\eta)(1+\xi)\\
N_8 &= \frac{1}{2}(1-\xi)(1-\eta)(1+\eta)
\end{aligned}
$$

The element strains are given by

$$\mathbf{e} = [e_{11}, e_{22}, 2e_{12}]^t = [e_{xx}, e_{yy}, \gamma_{xy}]^t, \tag{3.87}$$

where

$$e_{xx} = e_{11} = \frac{\partial u_1}{\partial x}, \quad e_{yy} = e_{22} = \frac{\partial u_2}{\partial y}, \quad \gamma_{xy} = 2e_{12} = \frac{\partial u_1}{\partial y} + \frac{\partial u_2}{\partial x}. \tag{3.88}$$

They involve the derivatives with respect to the global coordinates x and y. Using the chain rule, we have

$$\begin{vmatrix} \dfrac{\partial}{\partial \xi} \\ \dfrac{\partial}{\partial \eta} \end{vmatrix} = \begin{vmatrix} \dfrac{\partial x}{\partial \xi} & \dfrac{\partial y}{\partial \xi} \\ \dfrac{\partial x}{\partial \eta} & \dfrac{\partial y}{\partial \eta} \end{vmatrix} \begin{vmatrix} \dfrac{\partial}{\partial x} \\ \dfrac{\partial}{\partial y} \end{vmatrix}, \tag{3.89}$$

or in tensor notation, with $\mathbf{x} = [x, y]^t$, $\boldsymbol{\xi} = [\xi, \eta]^t$,

$$\frac{\partial}{\partial \boldsymbol{\xi}} = \mathbf{J}\frac{\partial}{\partial \mathbf{x}} \quad \text{or} \quad \frac{\partial}{\partial \mathbf{x}} = \mathbf{J}^{-1}\frac{\partial}{\partial \boldsymbol{\xi}}, \tag{3.90}$$

where $\mathbf{J}$ is the *Jacobian matrix* relating the global coordinate derivatives to the local coordinate derivatives, i.e.,

$$\mathbf{J} = \begin{vmatrix} \dfrac{\partial x}{\partial \xi} & \dfrac{\partial y}{\partial \xi} \\ \dfrac{\partial x}{\partial \eta} & \dfrac{\partial y}{\partial \eta} \end{vmatrix}, \quad \mathbf{J}^{-1} = \begin{vmatrix} \dfrac{\partial \xi}{\partial x} & \dfrac{\partial \eta}{\partial x} \\ \dfrac{\partial \xi}{\partial y} & \dfrac{\partial \eta}{\partial y} \end{vmatrix}$$

and $\mathbf{J}^{-1}$ is linked to $\mathbf{J}$ as

$$\mathbf{J} \triangleq \begin{vmatrix} J_{11} & J_{12} \\ J_{21} & J_{22} \end{vmatrix}, \quad \mathbf{J}^{-1} = \frac{1}{j}\begin{vmatrix} J_{22} & -J_{12} \\ -J_{21} & J_{11} \end{vmatrix}, \quad j \equiv \det(\mathbf{J}) = J_{11}J_{22} - J_{12}J_{21}. \tag{3.91}$$

Substituting Eq. (3.86) into Eq. (3.88), we have

$$\frac{\partial u_1}{\partial x} = N_{1\alpha,x}U_\alpha \quad \frac{\partial u_1}{\partial y} = N_{1\alpha,y}U_\alpha \quad \frac{\partial u_2}{\partial x} = N_{2\alpha,x}U_\alpha \quad \frac{\partial u_2}{\partial y} = N_{2\alpha,y}U_\alpha. \tag{3.92}$$

Hence

$$e_{ij} = B_{ij\alpha}U_\alpha, \tag{3.93}$$

with

$$B = \begin{vmatrix} N_{1,x} & 0 & N_{2,x} & 0 & \cdots & N_{8,x} & 0 \\ 0 & N_{1,y} & 0 & N_{2,y} & \cdots & 0 & N_{8,y} \\ N_{1,y} & N_{1,x} & N_{2,y} & N_{2,x} & \cdots & N_{8,y} & N_{8,x} \end{vmatrix}, \tag{3.94}$$

$$N_{\alpha,x} = (N_{\alpha,\xi}J_{22} - N_{\alpha,\eta}J_{12})/j, \tag{3.95}$$

$$N_{\alpha,y} = (-N_{\alpha,\xi}J_{21} + N_{\alpha,\eta}J_{11})/j. \tag{3.96}$$

Finite Element Formulation

Finite element formulation can be obtained by either weighted residual method or variational principle. Both methods will give same formulation for finite element methods.

Following the same procedure and using

$$t_{ij} = a_{ijkl}e_{kl} + b_{ijkl}\dot{e}_{kl}, \tag{3.97}$$

$$\hat{u}_i = N_{i\alpha}U_\alpha, \tag{3.98}$$

$$\hat{e}_{ij} = B_{ij\alpha}U_\alpha, \dot{\hat{e}}_{ij} = B_{ij\alpha}\dot{U}_\alpha, \tag{3.99}$$

the finite element formulation for viscoelastic solid can be obtained as

$$U_\beta \int_\Omega a_{ijkl} B_{kl\beta} B_{ij\alpha}\, \mathrm{d}V + \dot{U}_\beta \int_\Omega b_{ijkl} B_{kl\beta} B_{ij\alpha}\, \mathrm{d}V + \ddot{U}_\beta \int_\Omega \rho N_{i\beta} N_{i\alpha}\, \mathrm{d}V$$
$$= \int_{\Gamma_t} \bar{t}_i N_{i\alpha}\, \mathrm{d}S + \int_\Omega \rho f_i N_{i\alpha}\, \mathrm{d}V, \tag{3.100}$$

which can be rewritten in a general form of dynamic equation as

$$\mathbf{M}\ddot{\mathbf{U}} + \mathbf{C}\dot{\mathbf{U}} + \mathbf{K}\mathbf{U} = \mathbf{F}, \tag{3.101}$$

with the mass, damping, stiffness, and the external force matrices as

$$M_{\alpha\beta} = \int_V \rho \Phi_{i\beta}(\boldsymbol{x}) \Phi_{i\alpha}(\boldsymbol{x})\, \mathrm{d}V = M_{\beta\alpha}, \tag{3.102}$$

$$C_{\alpha\beta} = \int_V b_{ijkl} B_{ij\alpha}(\boldsymbol{x}) B_{kl\beta}(\boldsymbol{x})\, \mathrm{d}V, \tag{3.103}$$

$$K_{\alpha\beta} = \int_V a_{ijkl} B_{ij\alpha}(\boldsymbol{x}) B_{kl\beta}(\boldsymbol{x})\, \mathrm{d}V, \tag{3.104}$$

$$F_\alpha = \int_V \rho f_j \Phi_{j\alpha}(\boldsymbol{x})\, \mathrm{d}V + \int_{\Gamma_t} \bar{t}_j \Phi_{j\alpha}(\boldsymbol{x})\, \mathrm{d}S. \tag{3.105}$$

For static problems, ignoring the dynamic effect of $\ddot{\mathbf{U}}$ and $\dot{\mathbf{U}}$, Eq. (3.101) is then reduced to

$$\mathbf{K}\mathbf{U} = \mathbf{F}. \tag{3.106}$$

This simple form is the finite element equations of elastostatics.

Numerical Integration in Finite Element Method

Numerical integration is a very important step in finite element methods. The integrals, such as those in Eqs. (3.102–3.105), that we obtained previously can all be expressed in the form of

$$\int F(\xi, \eta)\, \mathrm{d}\xi\, \mathrm{d}\eta, \qquad \int F(\xi, \eta, \varsigma)\, \mathrm{d}\xi\, \mathrm{d}\eta\, \mathrm{d}\varsigma$$

for two- and three-dimensional problems, respectively. It was stated that those integrals are in practice evaluated numerically using

$$\int F(\xi, \eta)\, \mathrm{d}\xi\, \mathrm{d}\eta = \sum_{i,j} \alpha_{ij} F(\xi_i, \eta_j) + R_n, \tag{3.107}$$

$$\int F(\xi, \eta, \varsigma)\, \mathrm{d}\xi\, \mathrm{d}\eta\, \mathrm{d}\varsigma = \sum_{i,j,k} \alpha_{ijk} F(\xi_i, \eta_j, \varsigma_k) + R_n, \tag{3.108}$$

where the summations are extended over all i, j, and k as specified, the α_{ij} and α_{ijk} are weighting factors, $F(\xi_i, \eta_j)$ and $F(\xi_i, \eta_j, \varsigma_k)$ are the functions of $F(\xi, \eta)$ and $F(\xi, \eta, \varsigma)$ evaluated at the points specified in the arguments, respectively. The matrices R_n are error matrices which in practice are usually not evaluated.

The numerical integration of $\int_a^b F(\xi)\,d\xi$ is essentially based on passing a polynomial $\psi(\xi)$ through given values of $F(\xi)$ and then using $\int_a^b \psi(\xi)\,d\xi$ as an approximation to the interval from a to b to determine how well $\psi(\xi)$ approximates $F(\xi)$ and hence the error of the numerical integration.

A convenient way to obtain $\psi(\xi)$ is to use *Lagrangian interpolation*. The fundamental polynomials of *Lagrangian interpolation* is given as

$$l_j(\xi) = \frac{(\xi-\xi_0)(\xi-\xi_1)(\xi-\xi_2)\cdots(\xi-\xi_{j-1})(\xi-\xi_{j+1})\cdots(\xi-\xi_n)}{(\xi_j-\xi_0)(\xi_j-\xi_1)(\xi_j-\xi_2)\cdots(\xi_j-\xi_{j-1})(\xi_j-\xi_{j+1})\cdots(\xi_j-\xi_n)}. \tag{3.109}$$

It is seen that

$$l_j(\xi_i) = \delta_{ij}. \tag{3.110}$$

The polynomial $\psi(\xi)$ is then

$$\psi(\xi) = F_0 l_0(\xi) + F_1 l_1(\xi) + \cdots + F_n l_n(\xi). \tag{3.111}$$

The Newton–Cotes Formulas for One-Dimensional Integration

In Newton–Cotes integration, it is assumed that the sampling points of $F(\xi)$ are spaced at equal distance, and we have

$$\xi_0 = -1, \quad \xi_n = 1, \quad h = \frac{1}{n}, \tag{3.112}$$

where ξ_0 and ξ_n are the positions of the starting and ending points of a one-dimensional element, respectively and h is the spacing between sampling points.

Using Lagrangian interpolation to obtain $\psi(\xi)$ as an approximation to $F(\xi)$, it gives

$$\int_{-1}^{1} F(\xi)\,d\xi = \sum_{i=0}^{n}\left(\int_{-1}^{1} l_i(\xi)\,d\xi\right) F_i + R_n, \tag{3.113}$$

$$\int_{-1}^{1} F(\xi)\,d\xi = \sum_{i=0}^{n} C_i^n F_i + R_n, \tag{3.114}$$

where C_i^n are the Newton–Cotes constants for numerical integration with n sampling points. The cases $n = 1$ and $n = 2$ are the well-known *trapezoidal rule* and *Simpson formula*. The even formulas with $n = 2$ and $n = 4$ are used in practice. The error is of order $O(b^n)$ where b is the element size.

The Gauss Formulas for One-Dimensional Integration

A very important numerical integration approach in which both the positions of the sampling points and the weights have been optimized is the *Gauss quadrature*.

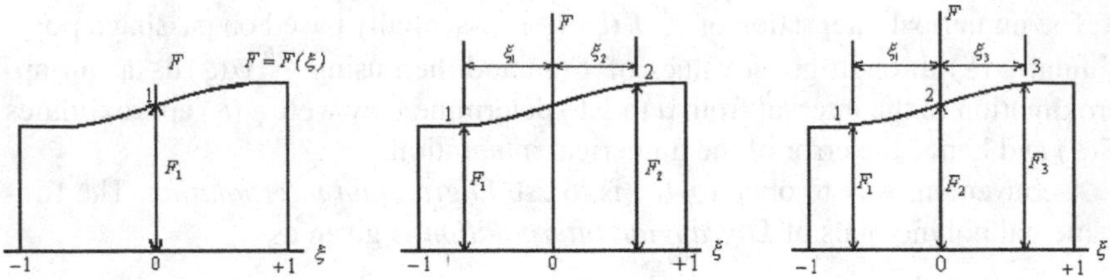

FIGURE 3.6. Gauss numerical integration for one, two, and three sampling points.

The basic assumption in Gauss numerical integration is that

$$\int_a^b F(\xi)\,d\xi = H_1 F(\xi_1) + H_2 F(\xi_2) + \cdots + H_n F(\xi_n) + R_n \approx \sum_{i=1}^{n} H_i F(\xi_i), \tag{3.115}$$

where both the weights H_i and the sampling point positions ξ_i are variables (cf. Fig. 3.6). If we assume a polynomial expression, it is easy to see that for n sampling points we have $2n$ unknowns, and hence a polynomial of degree $2n-1$ could be constructed and exactly integrated. The error thus is of order $O(h^{2n})$.

The simultaneous equations involved are difficult to solve. Fortunately, it is found that the solution can be obtained explicitly in terms of Legendre polynomials. Thus, this particular process is frequently known as the Gauss–Legendre quadrature. Table 3.1 summarizes the positions and weighting coefficients of *Gauss quadrature*.

TABLE 3.1. Abscissa and weight coefficients of the Gaussian quadrature formula (Zienkiewicz and Taylor, 1989)

$$\int_{-1}^{1} F(x)\,dx = \sum_{i=1}^{n} H_i F(a_i)$$

$\pm a$		H
	$n=1$	
0		2.000 000 000 000 000
	$n=2$	
0.577 350 269 189 626		1.000 000 000 000 000
	$n=3$	
0.774 596 669 241 483		0.555 555 555 555 556
0. 000 000 000 000 000		0.888 888 888 888 889
	$n=4$	
0.861 136 311 594 053		0.347 854 845 137 454
0.339 981 043 584 856		0.652 145 154 862 546
	$n=5$	
0.906 179 845 938 664		0.236 926 885 056 189
0.538 469 310 105 683		0.478 628 670 499 366
0.000 000 000 000 000		0.568 888 888 888 889
	$n=6$	
0.932 469 514 203 152		0.171 324 492 379 170
0.661 209 386 466 265		0.360 761 573 048 139
0.238 619 186 083 197		0.467 913 934 572 691

For the purpose of finite element analysis, the complex calculations are involved in determining the value of $F(\xi)$, the function to be integrated. Thus, the Gauss process, requiring the least number of such evaluations, is ideally suited, and has been used in most of finite element programs.

Two-Dimensional and Three-Dimensional Numerical Integrations

The two- and three-dimensional integrations can be obtained by applying the one-dimensional integration formulas successively in each direction. As in the analytical evaluation of multidimensional integrals, successively, the innermost integral is evaluated by keeping the variables corresponding to the other integrals constant. Therefore, we have

$$\int_{-1}^{1}\int_{-1}^{1} F(\xi,\eta)\,\mathrm{d}\xi\,\mathrm{d}\eta = \sum_{i=1}^{n} H_i \sum_{j=1}^{n} H_j F(\xi_i,\eta_j) = \sum_{i=1}^{n}\sum_{j=1}^{n} H_i H_j F(\xi_i,\eta_j), \tag{3.116}$$

$$\int_{-1}^{1}\int_{-1}^{1}\int_{-1}^{1} F(\xi,\eta,\varsigma)\,\mathrm{d}\xi\,\mathrm{d}\eta\,\mathrm{d}\varsigma = \sum_{i=1}^{n}\sum_{j=1}^{n}\sum_{k=1}^{n} H_i H_j H_k F(\xi_i,\eta_j,\varsigma_k). \tag{3.117}$$

It is noted that it is not necessary to have the number of integrating points to be same in each direction. Sometimes it may be of advantage to use different numbers in each direction of integration. It is also of interest to note that in fact the double or triple summation can be readily interpreted as a single one over $n \times n$ points for two-dimensional rectangle or $n \times n \times n$ points for a cube.

Required Order for Numerical Integration

In evaluating the matrices in Finite Element Formulation section, i.e., the mass matrix $\boldsymbol{M}$, stiffness matrix $\boldsymbol{K}$, damping matrix $\boldsymbol{C}$, and force vector $\boldsymbol{F}$, the choice of the order of numerical integration is most important because the cost of numerical integration can be quite significant for high-order integration; on the other hand using a different integration order, the results can be affected by a very large amount. It involves both the convergence and cost.

The integration order required to evaluate a specific element matrix accurately can be determined by studying the order of the function to be integrated. For example, the integral for the stiffness matrix,

$$K_{\alpha\beta} = \int_V a_{ijkl} B_{ij\alpha}(\boldsymbol{x}) B_{kl\beta}(\boldsymbol{x})\,\mathrm{d}V, \tag{3.118}$$

can be rewritten in terms of integral in the local coordinate of an element

$$K_{\alpha\beta} = \int a_{ijkl} B_{ij\alpha}(\boldsymbol{\xi}) B_{kl\beta}(\boldsymbol{\xi}) \det(\mathbf{J})\, d\boldsymbol{\xi}$$

$$= \int_{\xi=-1}^{1} \int_{\eta=-1}^{1} \int_{\varsigma=-1}^{1} a_{ijkl} B_{ij\alpha}(\xi, \eta, \varsigma) B_{kl\beta}(\xi, \eta, \varsigma) \det[\mathbf{J}(\xi, \eta, \varsigma)]\, d\xi\, d\eta\, d\varsigma, \tag{3.119}$$

where a_{ijkl} is the constant material property matrix, $B_{ij\alpha}(\boldsymbol{\xi})$ the strain–displacement relation in the local coordinate (ξ, η, ς) of an element, and $\det(\mathbf{J})$ the determinant of the Jacobian matrix transforming global to local coordinates, and the integration is performed over the element volume in local coordinate system. The matrix function F to be integrated is, therefore,

$$\mathbf{F} = \mathbf{B}^{\mathrm{T}} : \mathbf{a} : B \det(\mathbf{J}). \tag{3.120}$$

For a two-dimensional four-node rectangular element,

$$\mathbf{F} = f(\xi^2, \xi\eta, \eta^2), \tag{3.121}$$

hence, using two-point Gauss numerical integration is adequate; since for integration order n, the order of ξ and η integrated exactly is $(2n - 1)$ with Gaussian quadrature.

The recommended order for eight-node rectangular element is 3×3, whereas for 12-node rectangular element is 4×4. With these integration orders, the element matrices of geometrically undistorted elements can be evaluated exactly, whereas for geometrically distorted elements a sufficiently accurate approximation can be obtained unless the geometric distortions are extremely large.

To reduce the cost, the advantage of symmetry should be fully utilized. Finite element methods possess the advantage that symmetry in loading and geometry can reduce the problem to manageable proportions with appropriate boundary conditions.

Also, a concise presentation of relevant techniques is provided here; there is a rich and mature literature that could be referred to for a more detailed coverage of finite element methods and their various applications.

Problems

1. Reformulate the linear elastic problem by using penalty functions method.
2. For a two-dimensional linear elastic problem to have unique solution, what kind of boundary condition should be specified?
3. Consider the system of equations

$$\begin{vmatrix} 2 & -1 \\ -1 & 2 \end{vmatrix} \begin{vmatrix} U_1 \\ U_2 \end{vmatrix} = \begin{vmatrix} 10 \\ -1 \end{vmatrix}.$$

Use the Lagrange multiplier method and the penalty method to impose the condition $U_2 = 0$. Solve the equations and interpret the solution.

4. For thermoelastic solid, the constitutive equations for Cauchy stress tensor are usually written as

$$t_{ij} = -\beta_{ij}T + A_{ijkl}e_{kl},$$

where T is temperature deviation from the reference temperature (cf. Chapter 2); β_{ij} may be called the thermal expansion coefficients; and $-\beta_{ij}T$ are named as the thermal stresses. Show the finite element equations may be expressed as

$$\mathbf{M}\ddot{\mathbf{U}} + \mathbf{K}\mathbf{U} - \mathbf{P}\mathbf{T} = \mathbf{F},$$

where the expressions of **M**, **K**, and **F** are given in Eq. (3.30). What are the detailed expressions of **P** and **T**?

5. Following Problem 4, let the energy equation and the heat flux be written as

$$\rho^0\gamma\dot{T} + T^0\beta_{ij}\dot{e}_{ij} = -q_{i,i} + \rho^0 h$$
$$q_i = -H_{ij}T_{,j}$$

Show that another set of finite element equations may be obtained as

$$\mathbf{A}\dot{\mathbf{T}} + \mathbf{H}\mathbf{T} + T^0\mathbf{P}^t\dot{\mathbf{U}} = \mathbf{Q}$$

What are the detailed expressions of **A**, **H**, and **Q**?

6. Derive the **B** matrix for eight-node solid element.
7. Establish the Jacobian matrix **J** of the two-dimensional elements shown in the following.

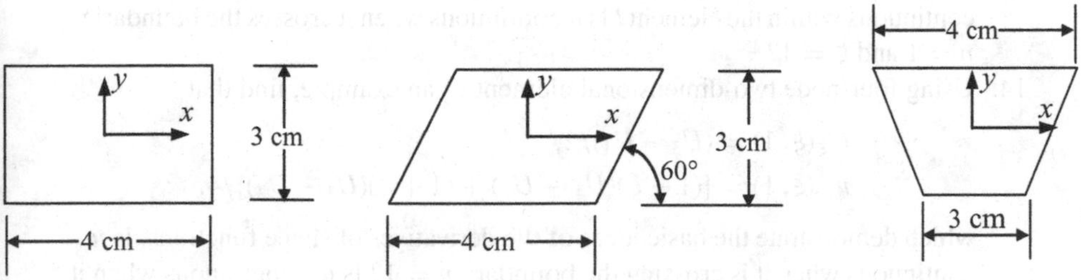

8. Use two-point Gauss quadrature to evaluate the integral $\int_0^4 (2^x - 2x + 4)\,dx$.
9. Verify that $\sum_{\alpha=1}^{m} N_\alpha = 1$ for all those mentioned linear, quadratic, and cubic two-dimensional and three-dimensional elements.
10. For four-node two-dimensional element, find that

$$\begin{aligned} N_{1,\xi} &= -(1-\eta)/4, & N_{1,\eta} &= -(1-\xi)/4 \\ N_{2,\xi} &= +(1-\eta)/4, & N_{2,\eta} &= -(1+\xi)/4 \\ N_{3,\xi} &= +(1+\eta)/4, & N_{3,\eta} &= +(1+\xi)/4 \\ N_{4,\xi} &= -(1+\eta)/4, & N_{4,\eta} &= +(1-\xi)/4 \end{aligned}$$

11. For eight-node two-dimensional element, find that

$$
\begin{aligned}
&N_{1,\xi} = (1-\eta)(2\xi+\eta)/4, \quad N_{1,\eta} = (1-\xi)(2\eta+\xi)/4\\
&N_{2,\xi} = (1-\eta)(2\xi-\eta)/4, \quad N_{2,\eta} = (1+\xi)(2\eta-\xi)/4\\
&N_{3,\xi} = (1+\eta)(2\xi+\eta)/4, \quad N_{3,\eta} = (1+\xi)(2\eta+\xi)/4\\
&N_{4,\xi} = (1+\eta)(2\xi-\eta)/4, \quad N_{4,\eta} = (1-\xi)(2\eta-\xi)/4\\
&N_{5,\xi} = -\xi(1-\eta), \quad N_{5,\eta} = -(1-\xi^2)/2\\
&N_{6,\xi} = (1-\eta^2)/2, \quad N_{6,\eta} = -\eta(1+\xi)\\
&N_{7,\xi} = -\xi(1+\eta), \quad N_{7,\eta} = (1-\xi^2)/2\\
&N_{8,\xi} = -(1-\eta^2)/2, \quad N_{8,\eta} = -\eta(1-\xi)
\end{aligned}
$$

12. Using four-node two-dimensional element as an example, show that the Jacobian matrix may be calculated as

$$
\mathbf{J} \equiv \begin{vmatrix} \dfrac{\partial x}{\partial \xi} & \dfrac{\partial y}{\partial \xi} \\ \dfrac{\partial x}{\partial \eta} & \dfrac{\partial y}{\partial \eta} \end{vmatrix} = \begin{vmatrix} N_{1,\xi} & N_{2,\xi} & N_{3,\xi} & N_{4,\xi} \\ N_{1,\eta} & N_{2,\eta} & N_{3,\eta} & N_{4,\eta} \end{vmatrix} \begin{vmatrix} X_1 & Y_1 \\ X_2 & Y_2 \\ X_3 & Y_3 \\ X_4 & Y_4 \end{vmatrix}
$$

where $\{X_i, Y_i\}$ $(i = 1, 2, 3, 4)$ are the coordinates of the four nodes of an element. In general, it is seen that the value of $\mathbf{J}$ depends on the position $\{\xi, \eta, \varsigma\}$ and the geometry of the element in question.

13. Using four-node two-dimensional element as an example, find that

$$
\begin{aligned}
u(0,0) &= (U_1 + U_2 + U_3 + U_4)/4,\\
u(\xi,1) &= \{(1+\xi)U_3 + (1-\xi)U_4\}/2,\\
u(1,\eta) &= \{(1-\eta)U_2 + (1+\eta)U_3\}/2,
\end{aligned}
$$

which demonstrate the basic ideas of shape functions. Is the function $u(\xi, \eta)$ continuous within the element? Is u continuous when it crosses the boundaries $\eta = 1$ and $\xi = 1$?

14. Using four-node two-dimensional element as an example, find that

$$
\begin{aligned}
u_{,\xi}(\xi,1) &= (U_3 - U_4)/2,\\
u_{,\eta}(\xi,1) &= \{(1-\xi)(U_4 - U_1) + (1+\xi)(U_3 - U_2)\}/4,
\end{aligned}
$$

which demonstrate the basic ideas of the derivatives of shape functions. Is $u_{,\xi}$ continuous when it is crossing the boundary $\eta = 1$? Is $u_{,\eta}$ continuous when it is crossing the boundary $\eta = 1$? So, strains and stresses are continuous within the element, but are they continuous when it crosses the element boundary?

4
An Overview on Meshless Methods and Their Applications

Meshless methods can be traced back to 1977 when Lucy (1977) and Gingold and Monaghan (1977) proposed a smooth particle hydrodynamics (SPH) method that was used for modeling astrophysical phenomena without boundaries, such as exploding stars and dust clouds. Extensive developments have been made in several varieties since then and with many different names: SPH (Monaghan, 1982, 1988, 1992), generalized finite difference method (Liszka and Orkisz, 1980), diffuse element method (Nayroles et al., 1992), particle in cell method (Sulsky et al., 1992), wavelet galerkin method (Qian and Weiss, 1993), reproducing kernel particle method (RKPM) (Liu et al., 1995a,b), element-free Galerkin (EFG) (Belytschko et al., 1994), partition of unity (PU) (Babuska and Melenk, 1995, 1996), Hp clouds (Duarte and Oden, 1995, 1996), finite point method (Onate et al., 1996a,b), free-mesh method (Yagawa and Furukawa, 2000), meshless local boundary integration equation method, meshless local Petrov–Galerkin method (MLPG) (Atluri and Zhu, 2000; Zhu, 1999), and multiscale methods (Liu et al., 1996a,b, 1997, 2000).

This chapter is to give an overview of the development of meshless methods, with emphasis on the approximation functions, the numerical implementation, and the applications.

Approximation Function

Meshless methods construct approximations entirely in terms of nodes. The *approximation function* is an essential feature of the method. A *weight function*, which plays an important role in the performance of the methods, is used in all varieties of meshless methods. The *compact support of weight functions*, also called the domain of influence of a node, gives a local character to the meshless methods. The weight function is nonzero in the domain of support and zero outside of the domain of support. The most commonly used supports are discs and rectangles, as shown in Fig. 4.1.

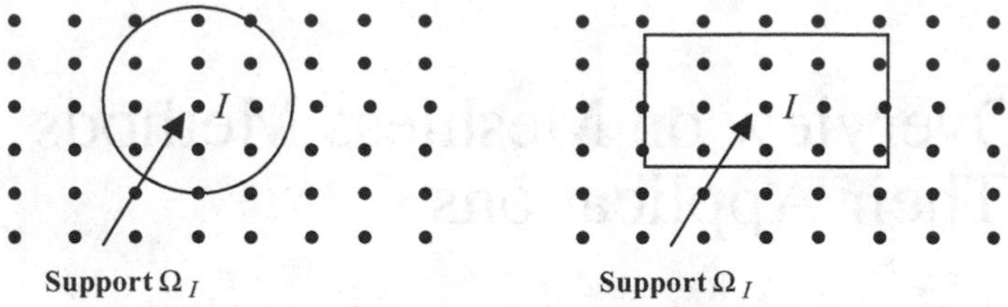

FIGURE 4.1. The commonly used supports of node I.

Smooth Particle Hydrodynamics Method

SPH method, the oldest meshless method, developed by Gingold and Monaghan in 1977, created a kernel approximation for a single function $u(\boldsymbol{x})$ in a domain Ω by

$$u^h(\boldsymbol{x}) = \int w(\boldsymbol{x} - \boldsymbol{y}, h)u(\boldsymbol{y})\, \mathrm{d}\Omega_y, \tag{4.1}$$

where $u^h(\boldsymbol{x})$ is the approximation, $w(\boldsymbol{x} - \boldsymbol{y}, h)$ a kernel or weight function, and h a measure of the size of the support.

The discrete form was obtained by numerical quadrature of the right-hand side in the following type:

$$u^h(\boldsymbol{x}) = \sum_I w(\boldsymbol{x} - \boldsymbol{x}_I)u_I \Delta V_I = \sum_I \phi_I(\boldsymbol{x})u_I, \tag{4.2}$$

where ΔV_I is the volume, for 3D, or area, for 2D, or length, for 1D, associated with node I, and $\phi_I(\boldsymbol{x}) = w(\boldsymbol{x} - \boldsymbol{x}_I)\Delta V_I$ the SPH shape function of the approximation. One difficulty in applying the above is the development of robust technique for assigning ΔV_I to each of the nodes. In SPH applications of hydrodynamics equations, the ambiguity associated with the definition of ΔV_I can be reduced by invoking the continuity equation.

For any numerical method to converge, it must be consistent and stable. Stability is associated with the quadrature of the Galerkin form and the character of the Galerkin procedure. Consistency inherently arises from the character of the approximation. The requirements depend on the order of the partial differential equations (PDEs) to be solved. For a PDE of order $2k$, solution by a Galerkin method requires consistency of order k, i.e., a constant field for the kth derivatives must be represented exactly as the discretization parameter h tends to zero. For a second order PDE, this implies that consistency is satisfied if constant first derivative can be represented exactly.

Consistency conditions are closely related to completeness and reproducing conditions. An approximation is complete if it provides a basis that can produce the function with an arbitrary order of accuracy. Any approximation that can exactly reproduce linear polynomials can reproduce any smooth function and its first derivative with arbitrary accuracy as the approximation is refined, and approximation that has linear consistency also has linear completeness. Reproducing condition refers to the ability to reproduce a function if the nodal values are set by the function. Therefore, the ability to reproduce nth order polynomials is equivalent to nth order consistency. The order of the consistency of an approximation is also called the order of the polynomial that can be exactly represented, and consistency conditions are often expressed in terms of the order of the polynomial that can be exactly represented.

It can be shown that the linear consistency condition does not hold for both uniform and nonuniform meshes in SPH.

Reproducing Kernel Particle Method

Along the same line of development as SPH, Liu and coworkers developed the RKPM and proposed a correction function for kernels in both the discrete and continuous cases. The reproducing kernel approximation of $u(\boldsymbol{x})$ is given by

$$u^h(\boldsymbol{x}) = \int C(\boldsymbol{x}, \boldsymbol{x}-\boldsymbol{y})\Phi_\alpha(\boldsymbol{x}-\boldsymbol{y})u(\boldsymbol{y})\,\mathrm{d}\Omega_y, \tag{4.3}$$

where $C(\boldsymbol{x}, \boldsymbol{x}-\boldsymbol{y})$ is called the correction function which is obtained by imposing the reproducing conditions, i.e., the reproducing equation should exactly reproduce polynomials and can be expressed by a linear combination of polynomial basis functions; α is the dilation parameter of the kernel function $\Phi_\alpha(\boldsymbol{x}-\boldsymbol{y})$. By performing the numerical integration, the following discrete form can be obtained:

$$u^h(\boldsymbol{x}) = \sum_I \bar{\Phi}_\alpha(\boldsymbol{x}, \boldsymbol{x}-\boldsymbol{x}_I)u(\boldsymbol{x}_I)\Delta V_I = \sum_I \phi_I(\boldsymbol{x})u_I. \tag{4.4}$$

This approach is motivated by the theory of wavelets in which a function is represented by a combination of the dilation and translation of a single wavelet, which is a window function. Based on the reproducing kernel particles and wavelets, the RKPM was further elaborated in the frequency domain and a multiple scale kernel particle method has been presented (Liu et al., 1995b, 1996a,b, 1997). This method permits the response of a system to be separated into different scales, with wave numbers corresponding to spatial scales and/or frequencies corresponding to temporal scales, and the response of each scale can be examined separately.

Moving Least Square Approximation

Nayroles et al. (1992) introduced a spatial discretization in the numerical solution of boundary value problems. In their method, only a set of nodes and a boundary description are needed to develop the Galerkin equations. The interpolants are polynomials that are fit to the nodal values by a least square approximation. The approximation was not recognized as moving least squares (MLSs), referred as "diffuse elements," and the method was viewed as a generalization of the finite element method (FEM).

Belytschko et al. (1994) recognized this spatial discretization as MLSs, refined and modified this method, and called it EFG. In MLS, the interpolants of the function $u(\boldsymbol{x})$ were defined in the domain Ω by

$$u^h(\boldsymbol{x}) = \sum_{i=1}^{m} p_i(\boldsymbol{x})a_i(\boldsymbol{x}) \equiv \boldsymbol{p}^T(\boldsymbol{x})\boldsymbol{a}(\boldsymbol{x}), \quad (4.5)$$

where m is the number of terms in the basis, $p_i(\boldsymbol{x})$ the monomial basis functions, and $a_i(\boldsymbol{x})$ the coefficients which are functions of the spatial coordinates $\boldsymbol{x}$. Examples of commonly used bases are the linear basis and the quadratic basis. The coefficients $a_i(\boldsymbol{x})$ are obtained by performing a weighted least square fit for the local approximation, which is obtained by minimizing the difference between the local approximation and the function, with

$$\begin{aligned} J &= \sum_I w(\boldsymbol{x}-\boldsymbol{x}_I)[u^h(\boldsymbol{x},\boldsymbol{x}_I) - u(\boldsymbol{x}_I)]^2, \\ &= \sum_I w(\boldsymbol{x}-\boldsymbol{x}_I)\left[\sum_i p_i(\boldsymbol{x}_I)a_i(\boldsymbol{x}) - u(\boldsymbol{x}_I)\right]^2, \\ &= (\boldsymbol{Pa}-\boldsymbol{u})^T\boldsymbol{W}(\boldsymbol{x})(\boldsymbol{Pa}-\boldsymbol{u}), \qquad (4.6) \\ \frac{\partial J}{\partial \boldsymbol{a}} &= \boldsymbol{A}(\boldsymbol{x})\boldsymbol{a}(\boldsymbol{x}) - \boldsymbol{B}(\boldsymbol{x})u = 0. \qquad (4.7) \end{aligned}$$

The approximation $u^h(\boldsymbol{x})$ can then be expressed in the form of shape functions and nodal values as

$$u^h(\boldsymbol{x}) = \sum_I \phi_I^k(\boldsymbol{x})u_I. \quad (4.8)$$

The consistency of order k of the MLS approximations can be satisfied if the basis is complete in the polynomials of order k. In fact, any function, which appears in the basis, can be reproduced exactly by an MLS approximation.

Partition of Unity Methods

Babuska and Melenk (1995, 1996) and Duarte and Oden (1995, 1996) have shown that the MLS functions constitute a PU, and the methods based on MLS are specific instances of partitions of unity. Based on this viewpoint, they did powerful extensions of these approximations and proposed new approximations.

Babuska and Melenk (1995, 1996) introduced the following approximation form

$$u^h(\boldsymbol{x}) = \sum_{I=1} \phi_I^0(\boldsymbol{x})(a_{0I} + a_{1I}x + \cdots + a_{kI}x^k + b_{1I}\sinh nx + b_{2I}\cosh nx), \quad (4.9)$$

where $\phi^0(x)$ is the Shepard function, or the zeroth-order approximation, which produces the PU and hence the compact support of the approximation. The coefficients a_{kI}, b_{1I}, and b_{2I} are the unknowns of the approximation and can be determined by a Galerkin or collocation procedure. The degree of consistency depends on the number of terms x^k.

Duarte and Oden (1995, 1996) used the PU concept in a more general manner. Their approximation is

$$u^h(\boldsymbol{x}) = \sum_I \phi_I^k(\boldsymbol{x}) \left(u_I + \sum_{i=1}^{m} b_{iI} q_i(\boldsymbol{x}) \right). \tag{4.10}$$

Here, $q_i(\boldsymbol{x})$ can be a monomial basis of any order greater than k and can be either higher-order monomials or enhancement functions, called extrinsic base. These facilitate the hp adaptivity.

The concept of the PU provides a rational method for constructing localized approximations to global functions with a greater degree of flexibility. Based on the concept, Krysl and Belytschko (2000) presented an approach to construct basis functions of a linear-precision PU for unstructured meshless methods. In their approach, they used the Shepard functions as the PU and designed the nodal functions so that the degrees of freedom are exclusively the values of the approximation function at nodes. The benefits of this approach include better conditioning of the discrete equations and easier handling of essential boundary conditions in applications to PDEs.

Other Meshless Methods

Liszka et al. (1980, 1996) took another path in the evolution of meshless methods with a "generalized finite difference method." Although it is cast as a finite difference method for arbitrary grids, it also used a least square fit, for which the parameters are the partial derivatives of the function at nodes. This method is applicable to arbitrary domains and employs only a scattered set of nodes to build approximation solutions to boundary value problems. It has been shown that the method takes a character that closely resembles MLS and partitions of unity (Liszka et al., 1996).

Yagawa et al. (1996, 2000) proposed a seamless FEM, namely the "free-mesh method." In their approach, the elements were created around each node only in a local manner after a set of nodes were provided; thereby, the processes from the creation of the local elements to the construction of the local equations were conducted on a node-by-node basis. The main advantage of their method in comparison with other meshless method is its reliability of solution as the construction of global linear equations is based on FEM, accuracy of which has been studied for decades.

Sulsky et al. (1992) described a "particle-in-cell method" within the framework of conventional finite element technology. In their approach, particles were interpreted as material points that were followed through the complete loading process. A fixed Eulerian grid provided the means for determining a spatial gradient. With the use of maps between the material points and the grid, the advantages of both Eulerian and Lagrangian schemes could be utilized so that the mesh tangling is avoided while the material variables are tracked through the complete deformation history.

Onate et al. (1996a) presented a meshless method called the "finite point method" for solving convection–diffusion and fluid flow type problems. The approach is based on a weighted least square interpolation of point data and point collocation for evaluating the approximation integrals. Improvement to this method was given (Onate et al., 1996b) to the stabilization of the convective terms and the Neumann boundary condition. Numerical examples have shown that the method for solution of adjoint and nonself adjoint equations typical of convective–diffusive transport and also to the analysis of a compressible fluid mechanics problem worked well.

"Natural Neighbor Interpolation" (Sibson, 1981) is a multivariate scattered data interpolation method that has primarily been used in geophysical modeling. This natural neighbor interpolant relies on the concept of the Voronoi diagram and Delaunay triangulations, and has very interesting features, such as its strictly interpolant character and its ability to exactly interpolate piecewise linear boundary conditions. The application of natural neighbor coordinates to the numerical solution of PDEs was carried out by Traversoni (1994) and Braun and Sambridge (1995). The latter researchers coined the name "Natural Element Method" (NEM) to refer to its numerical implementation. Cueto et al. (2000) have imposed essential boundary conditions in NEM by means of density-scaled α-shapes. Sukumar et al. (1998, 1999, 2001) have systematically used NEM to solve solid mechanics problems.

Common Feature of the Approximations

It has been shown that the kernel methods, MLS methods, and PU methods share many feature of the same framework (Belytschko et al., 1996c). A discrete kernel approximation that is consistent must be identical to the related MLS approximation. Replacing the discrete sum in the MLS approximation by an integral leads to an approximation similar to that of SPH (Belytschko et al., 1994). Liu et al. (1996a) have identified a similar correspondence in a different way and have called the difference between the SPH approximation and a generalized reproducing kernel as the correction function; they assert that this correction function is essential for accuracy near boundaries. Duarte and Oden (1995) pointed out that any MLSs approximation could serve as a PU.

Numerical Implementation

Collocation Method

The collocation method was employed in the SPH for the discretization. The discrete equations of approximation were obtained by enforcing the approximation equation on a set of interior nodes. The equations obtained are just a set of algebraic equations in the unknown variables. This is obviously a simple and fast method, but it has been reported to suffer from instability.

Dyka (1994), Dyka et al. (1997), and Randles and Libersky (1996) have proposed stabilization by means of stress particles. Swegle et al. (1995) have shown the origin of the tensile instability through a dispersion analysis of the linearized equations and proposed a viscosity to stabilize it. Johnson and Beissel (1996) have proposed a method for improving the strain calculation. In a recent article by Randles et al. (1999), two sets of points were created for domain discretization; one carries velocity and another carries stress. It was reported that this treatment could improve the accuracy and reduce spurious oscillations in the SPH.

Galerkin Method with Quadrature Integration Scheme

Discretization by the Galerkin method requires a weak form or a variational principle. EFG (Belytschko et al., 1994), Hp clouds (Duarte and Oden, 1995, 1996), PU (Babuska and Melenk, 1995, 1996), and RKPM (Liu et al., 1995b) use the Galerkin method to obtain the discrete equations of the approximation, Lagrangian multipliers to enforce the essential boundary condition, and a quadrature scheme (cell quadrature or element quadrature) to evaluate the integrals. It has been reported that this procedure did not exhibit any volumetric locking, and the rate of convergence could exceed that of FEM significantly (Belytschko et al., 1994). The disadvantage of this method seems to be that the resulting method is not truly meshless, and its cost is too high. It was reported that the computational cost of explicit EFG exceeds a low-order FEM by a factor of 4–10 (Belytschko et al., 1996a,b,c, 2000).

Improvements were introduced by Lu et al. (1994) by employing an orthogonal polynomial basis to reduce the computational effort in constructing the MLS approximation and by replacing the Lagrangian multipliers with their physical counterparts to modify the variational principle, which lead to a banded, positive-definite stiffness matrix. While this modified formulation is not as accurate as the original one, it is suggested that by using a finer grid, equivalent accuracy could be achieved with less computational effort.

Nodal Integration of Galerkin Method

Spatial integration of the Galerkin method was achieved by evaluating the integrals of the weak form only at the nodes (Beissel and Belytschko, 1996). There was no need of cell structure or background mesh in this approach. It is a truly meshless method and is faster. However, it results in a spatial instability. Beissel and Belytschko (1996) proposed a stabilization procedure for the application of nodal integration to elastostatics by adding the square of the residual of equilibrium equation to the potential energy functional. It has been shown that the unstabilized EFG method with nodal integration manifests near-singular modes in many cases and that the stabilized EFG method eliminated these modes and achieved a reasonable rate of convergence.

A strain smoothing stabilization was introduced to compute nodal strain by a divergence counterpart of a spatial averaging of strain for nodal integration of Galerkin (Chen et al., 2001a,b). It was reported that this could avoid evaluating derivatives of meshless shape functions at nodes and thus eliminate spurious modes.

A unified stability analysis of meshless methods with Eulerian and Lagrangian kernels was presented by Belytschko et al. (2000). The stability properties of EFG and SPH under different quadrature scheme were investigated using Fourier analysis. Three types of instabilities were identified in one-dimensional problems: (1) an instability which occurs due to rank deficiency of the discretization of the divergence and makes the equilibrium equations singular; (2) a tensile instability which occurs when the stress is tensile and the second derivative of the kernel is large enough; and (3) an instability under compressive stress which also occurs in the continuum equations. It was suggested that the best approach to stabilize particle discretization of solids and fluids is to use Lagrangian kernels with stress points.

Local Boundary Integral Equation Method and Local Petrov–Galerkin Method

Zhu (1999) and Atluri and Zhu (2000) proposed two kinds of meshless methods: meshless local boundary integral equation (MLBIE) method and MLPG method. Both methods used MLS approximation to interpolate the solution variables, whereas the MLBIE method used a local boundary integral equation formulation, and the MLPG employed a local symmetric weak form. Integrals in both methods were evaluated over regularly shaped domains and their boundaries. There is no need for a background mesh, and they called the methods truly meshless methods.

Imposition of Essential Boundary Conditions

The meshless approximation does not pass through the nodal parameter values. As a consequence, the imposition of boundary conditions on the dependent variable is one of the main difficulties in the implementation of meshless methods. There have been several different approaches to this problem:

1. Lagrangian multiplier approaches (Aluru, 1999; Babuska and Melenk, 1995; Belytschko et al., 1994; Duarte and Oden, 1995; Liu et al., 1995a).
2. Modified variational principles (Lu et al., 1994).
3. Penalty methods (Atluri and Zhu, 2000; Kim et al., 2000).
4. Perturbed Lagrangian (Chu and Moran, 1995).
5. Point collocation method (Wagner and Liu, 2000).
6. Coupling to finite element (Attaway et al., 1994; Belytschko et al., 1995; Johnson, 1994; Liu and Chen, 1995).
7. NEMs (Cueto et al., 2000).

Among the above-mentioned methods, the Lagrangian multiplier method is the most accurate one for imposing Dirichlet boundary conditions; however, it is the most expensive one since the discrete equations for a linear self-adjoint PDE are no longer positive definite nor banded (Belytschko et al., 1996c). It was suggested that, since meshless methods have been not as fast as FEMs, it is advantageous to use meshless method only in those subdomains where their greater versatility is needed; e.g., problems with moving discontinuities, and use finite element model for the rest (Belytschko et al., 1996c).

Applications

Large Deformation Analysis

Finite element formulations dealing with geometric and material nonlinearities have been well developed, and a significant amount of work has been accomplished in large deformation analysis. However, the underlying structure of the FEM that originates from their reliance on a mesh is not well suited to the treatment of extreme mesh distortion. Meshless methods require no explicit mesh in computation and therefore avoid mesh distortion difficulties in large deformation analysis.

A 3D explicit EFG has been formulated and applied to the Taylor bar, the classical benchmark for nonlinear elastic–plastic computations, by Belytschko et al. (1996c). Both normal impact and oblique impact were simulated. Numerical examples also included fluid sloshing. The accuracy and cost were compared with explicit FE DYNA3D models. It was shown that the EFG model is more accurate than the FE model with the same number of degrees of freedom; however, it is more expensive than FE model. Simulations (Belytschko et al., 1996c) also indicated that the EFG was free of volumetric locking in incompressible materials.

Liu et al. (1996a) simulated an elastic rubber ring impacting with a rigid wall by RKPM. Jun et al. (1998) formulated an explicit RKPM and applied it to simulations of nonlinear elastic materials. Several numerical examples, namely bending of a rubber beam, necking of a circular bar, and an elastic–plastic aluminum bar impacting with a rigid wall, were presented by Chen et al. (1996) using RKPM. It was found that, compared to FEM, RKPM could better handle large deformation without any special numerical treatment. Also, there was an absence of volumetric locking for RKPM in dealing with incompressible materials under large deformation.

Li et al. (2000) simulated shear band formations in an elasto-viscoplastic material using reproducing kernel interpolants in a displacement-based explicit formulation. It was shown that the numerical solutions obtained are insensitive to the orientation of the particle distributions if the local particle distribution is quasi-uniform. Although mesh-alignment sensitivity could not be completely eliminated, especially in the case where the particle density is not quasi-uniform, the situations have been drastically improved compared with those of the FEM. The advantages of using meshless methods to simulate shear band formation were concluded to be its ability to avoid volumetric locking, the nonlocal feature of its approximation, and a favorable environment for hp-adaptive refinement.

Fracture and Crack Propagation

Meshless methods differ from the FEM in that the discretization is achieved by a model that consists only of nodes and a description of the internal and external boundaries and interfaces, like cracks, of the model. The connectivity in terms of node interaction may vary with time and space. The interaction of the nodes can be changed by the evolution of the model. Those flexible meshless features considerably simplify the modeling of fracture, free surface, and crack propagation. Meshless methods have much greater advantage in the treatment of evolving moving discontinuities such as crack growth than the FEM and the finite difference method.

The EFG method has been shown to be well suited to simulate arbitrary dynamic crack propagation by Belytschko et al. (1996a,b). In those simulations, a constant crack-tip velocity was assumed. An extension and application to process zone models by Belytschko et al. (2000) was presented with mixed-mode dynamic crack propagation in concrete. The method is suited for arbitrary crack growth with respect to crack speed and direction. The formulation included a model of the fracture process zone to characterize fracture processes in concrete and other cement-based materials. The computational results were shown to agree well with the experimental results. It was reported that the running time for EFG were approximately five times longer than that of an equivalent FE approach. If viewed in terms of the costs involved to solve this problem with interactive remeshings by a user, these computer costs are quite modest.

Ponthot and Belytschko (1998) presented an Arbitrary Lagrangian–Eulerian (ALE) formulation for EFG method and applied it to the crack propagation problems. The ALE formulation allows continuous relocation of nodes on the computational domain and keeps the density of nodes dynamically high through the evolving region where it is needed and at a rather small cost. Schwer et al. (2000) applied EFG to simulations of concrete failure in dynamic uniaxial tension tests. A set of EFG simulations were performed with a single crack and with multiple cracks in an effort to assess the effect on the computed strains of the presence and location of cracks in the rod. Numerical results and their comparisons to laboratory measurements provide some insight into the subtleties in the experimental strain histories and improve the understanding of the dynamic tensile failure of concrete.

Thin-Shell Structure

The numerical simulation of linear/nonlinear thin-shell structures has been a challenge in applied mechanics. Its applications cover many engineering areas such as metal forming, vehicle crashworthiness, vessel liability, etc., and has been a research area for a century. Constructing C^1 finite elements for shells of general shape has been addressed by many researchers and methodologies which circumvent the continuity requirement seem to have become predominant in recent years. From this viewpoint, meshless methods based on MLS approximations are very attractive for shell structures because C^1 continuity requirement can be easily met by its approximation.

Krysl and Belytschko applied EFG to the analysis of thin Kirchhoff plates (Krysl and Belytschko, 1996a) and to the analysis of thin shells (Krysl and Belytschko, 1996b) with C^1 approximations. They used background quadrilateral elements for the purpose of integration and Lagrangian multipliers to enforce the essential boundary conditions. Both applications were successful, as demonstrated by the numerical examples. The shell theory used was geometrically exact and could be applied to deep shells. EFG offers potential to those boundary value problems that require high continuity in the trial functions, and the Kirchoff shell theory is one of them. It was found that membrane locking, which is due to a different approximation order for transverse and membrane displacements, was removed by using larger domains of influence with the quadratic basis, and the locking was removed completely by using a quadratic polynomial basis.

Donning and Liu (1998) presented an approach to analyze moderately thin and thick structures using Mindlin–Reissner theory. They used an unmodified displacement-based Galerkin method and a uniform discretization. It was reported that shear and membrane lockings were completely eliminated pointwise at the interpolant level using cardinal splines, and the method worked well for coarse discretization.

Instead of using thin-shell theory, Li et al. (2000) performed a large deformation analysis of thin-shell structures with a 3D continuum using a meshless method. They used window function-based meshless interpolants to directly simulate large deformation of thin-shell structures. Numerical results have shown that the approach was viable in 3D direct simulation of thin-shell structures that are undergoing extremely large deformations. The main advantage of this approach is its simplicity in both formulation and implementation.

Shape Design Sensitivity Analysis and Optimization

The difficulty of shape design optimization problems arises from the fact that the geometry of the structure is the design variable. This means that the analysis model associated with a structure must be changed in a process of optimization. Normally, it requires several times of remeshing from an initial shape to an optimum shape with a mesh-based FEM, and it is essential for the finite element meshes to vary smoothly during the iterations. In addition, highly skewed finite element meshes can generate difficulties in computation of design sensitivities and lead to instability of the overall optimization process.

Grindeanu et al. (1998, 1999) performed a design sensitivity analysis of hyperelastic structures and a shape design optimization of hyperelastic structures by using meshless RKPM. In their papers, the Rivilin energy density function was employed to describe the hyperelastic structural behavior and the Lagrangian multiplier method was used for imposing the essential boundary conditions. The numerical results have shown that shape design sensitivity analysis and optimization using meshless methods can eliminate mesh distortion problems that occurs in finite element analysis using ABAQUS. Kim et al. (2000) performed shape design sensitivity analysis and optimization for a contact problem with friction. It also benefits from the fact that solution by the meshless method is much less sensitive to the mesh distortion. The accuracy of their numerical results has been compared with the results of FEMs, and good agreement has been obtained.

Micromechanics

Classical continuum physics is based on the fundamental assumption that all balance laws are valid for every part of the body, however small it may be, and that the state of the body at any material point is influenced only by the infinitesimal neighborhood about that point. The first of these assumptions eliminates the long-range effect of loads on the motion and the evolution of the state of the body, and the second ignores the effect of long-range interatomic interactions. This implies a certain intrinsic limitation (long wavelength limit) since the cohesive forces in real materials have a finite or even infinite range, and nonlocality is an intrinsic aspect of certain material phenomena.

Generally any problem that requires the solution of integro-differential equations can be said to be nonlocal in character. Eringen (1966) proposed a nonlocal elastic theory in the spirit of the well-accepted classical continuum mechanics with certain modifications. Edelen (1969) and Edelen and Laws (1971) proposed a nonlocal elastic theory in the tradition of Gibbsian thermodynamics employing a variational principle. Both approaches turned out to give similar results for elastic solids (Eringen and Edelen, 1972). Later, further developments in this area are almost within the framework of those two theories. The strain gradient theories (Aifantis, 1999; Fleck et al., 1994), which are reported to be able to explain many size-dependent phenomena and effective in dealing with plastic instabilities at the mesoscale (dislocation patterning) or macroscale (shear localization), can be considered as successive approximation to the nonlocal integral theory.

Meshless method is a computational tool. However, meshless approximations possess intrinsic nonlocal properties that make it easy and natural to incorporate with nonlocal theory. Moreover, the meshless method is a particle method, which is closer to the nature of the discrete atomistic model.

Liu et al. (2000) presented a multiscale method. The multiple field based on a 1D gradient plasticity theory with material length scale was proposed to remove the mesh dependency difficulty in softening/localization problems. Numerical results have shown that, in conjunction with the strain gradient theory, the multiple field RKPM could be applied for simulating strain localization problems. Chen et al. (2000) demonstrated that a length scale could be directly incorporated into a meshless approximation to regularize problems with material instabilities. Two types of implicit length scale implementations were incorporated: one in the displacement approximation and the other in the strain approximation. The numerical results were shown to match with the results of a gradient regularization.

Chen et al. (2001a,b, 2002a,b) developed a meshless method of nonlocal field theory. They suggested that both the meshless method and the nonlocal field theory could benefit from this incorporation. The nonlocal kernel in the constitutive theory brings the influence of strains at distant points $\boldsymbol{x}'$ to the stresses at $\boldsymbol{x}$. Meshless methods share this essential feature by nonlocally constructing the approximation in the domain of influence. The incorporation of meshless particle method and nonlocal constitutive theory takes the effect of long-range material interactions into consideration and brings the length scales into the picture. Numerical examples in their paper have proved its effectiveness in small length scale problems through the stress analyses of a crack problem and a beam problem in which the size effect is demonstrated.

Hoover and Hoover (1993) suggested a potential for the meshless particle method to apply to a variety of hybrid atomistic continuum problems in material science. They described the applications of the particle-based continuum technique from the closely related standpoint of nonequilibrium molecular dynamics. They pointed out that the logical and computational structure of SPH closely resembles that of ordinary molecular dynamics, although additional state variables are required. The dynamics, the treatment of boundary conditions, and the analysis of chaotic instability are also similar. They also suggested that the meshless particle approach should prove particularly valuable in constructing hybrid methods bridging the gap between the atomistic and the continuum views and in characterizing the fluctuations that underlie continuum flows.

Problems

1. Check the consistency of MLS approximation.
2. Make a literature search on
   (a) Stability analysis of meshless methods.
   (b) Treatment of essential boundary conditions.
   (c) Applications on microscopic physical problems.

5 Procedures of Meshless Methods

This chapter presents the procedures of meshless methods. A good understanding of the procedures of meshless methods would enrich our confidence in using and further developing meshless methods. For this purpose, certain procedures are described and discussed in detail.

Construction of the Approximation

It has been shown that the kernel methods, moving least-square methods, and partition of unity methods share many features of the same framework. A discrete kernel approximation that is consistent must be identical to the related moving least-square approximation. We are therefore going into some details of the moving least-square technique to show the procedures of constructing the approximation.

In moving least-square approximation, the approximation $\hat{u}(\mathbf{x})$ of a scalar-valued function, $u(\mathbf{x})$, is represented by the inner product of a vector of the polynomial basis, $\mathbf{p}(\mathbf{x})$, and a vector of the coefficients, $\mathbf{a}(\mathbf{x})$, as:

$$\hat{u}(\mathbf{x}) = \sum_{i=1}^{m} p_i(\mathbf{x})a_i(\mathbf{x}) \equiv \mathbf{p}^t(\mathbf{x})\mathbf{a}(\mathbf{x}) \equiv \mathbf{p}(\mathbf{x}) \cdot \mathbf{a}(\mathbf{x}), \tag{5.1}$$

where m is the number of terms in the polynomial basis; $p_i(\mathbf{x})$ $(i = 1, 2, 3, \ldots, m)$ are the monomial basis functions; $\mathbf{p}' = \mathbf{p}^t$ is the transpose of $\mathbf{p}$; and $a_i(\mathbf{x})$ are their corresponding coefficients that are functions of the spatial coordinates $\mathbf{x}$. Examples of the commonly used bases are the polynomial bases:

$$\mathbf{p}'(\mathbf{x}; 1, 1) = \{1, x\}, \tag{5.2}$$

$$\mathbf{p}'(\mathbf{x}; 1, 2) = \{1, x, y\}, \tag{5.3}$$

$$\mathbf{p}'(\mathbf{x}; 2, 2) = \{1, x, y, x^2, y^2, xy\}, \tag{5.4}$$

$$\mathbf{p}'(\mathbf{x}; 3, 2) = \{1, x, y, x^2, y^2, xy, x^3, y^3, x^2y, xy^2\}, \tag{5.5}$$

$$\mathbf{p}'(\mathbf{x}; 3, 3) = \{1, x, y, z, x^2, y^2, z^2, xy, yz, zx, x^3, y^3, z^3, x^2y, xy^2, y^2z, yz^2, z^2x, zx^2, xyz\}, \tag{5.6}$$

where $\mathbf{p}(\mathbf{x}; k, n)$ are the kth order n-dimensional polynomial basis functions of $\mathbf{x}$. Apparently, the number of coefficients, m, depends on the order of polynomial and dimensions. In order to determine $\mathbf{a}(\mathbf{x})$, a weighted discrete error norm can be constructed as

$$J(\mathbf{x}) = \sum_{I=1}^{n} W_I(\mathbf{x})[\mathbf{p}(\mathbf{x}_I) \cdot \mathbf{a}(\mathbf{x}) - U_I]^2, \tag{5.7}$$

where U_I are the data of u at the Ith node, x_I is the coordinate of the Ith node, $\mathbf{x}$ the coordinate of a generic sampling point, n the number of nodes of which the domain of influence (support) covers $\mathbf{x}$, the weight function associated with the Ith node is given by

$$W_I(\mathbf{x}) = W(\mathbf{x}, \mathbf{x}_I, \rho_I), \tag{5.8}$$

and ρ_I is the radius of the support of the Ith node. Then the error norm is minimized with respect to $\mathbf{a}(\mathbf{x})$, i.e.,

$$\frac{\partial J(\mathbf{x})}{\partial a(\mathbf{x})} = 0, \tag{5.9}$$

which leads to the following set of linear equations for $\mathbf{a}(\mathbf{x})$

$$\mathbf{A}(\mathbf{x})\mathbf{a}(\mathbf{x}) = \mathbf{B}(\mathbf{x})\mathbf{U} \tag{5.10}$$

where

$$\mathbf{U}^t = \{U_1, U_2, \ldots, U_n\}, \tag{5.11}$$

$$\mathbf{A} \equiv \sum_{I=1}^{n} W_I(\mathbf{x})\mathbf{p}(\mathbf{x}_I)\mathbf{p}^t(\mathbf{x}_I), \tag{5.12}$$

$$\mathbf{B} \equiv \{W_1(x)\boldsymbol{p}(x_1), W_2(x)\boldsymbol{p}(x_2), \ldots, W_n(x)\boldsymbol{p}(x_n)\}. \tag{5.13}$$

The vector of coefficients, $\mathbf{a}(\mathbf{x})$, can be solved as

$$\mathbf{a}(x) = \mathbf{A}^{-1}(x)\mathbf{B}(x)\mathbf{U}, \tag{5.14}$$

provided that the matrix $\mathbf{A}$ is nonsingular for every sampling point $\mathbf{x}$. Now the approximation of $u(\mathbf{x})$ can be expressed as

$$u \cong \hat{u} = \mathbf{p}'(\mathbf{x})\mathbf{A}^{-1}(\mathbf{x})\mathbf{B}(\mathbf{x})\mathbf{U} \equiv \boldsymbol{\Phi}(\mathbf{x})\mathbf{U}, \tag{5.15}$$

which takes the form of an inner product between vectors of shape functions, $\boldsymbol{\Phi}$, and nodal values, $\mathbf{U}$, as in the finite element method. The shape function and its derivative can be obtained as

$$\boldsymbol{\Phi}(\mathbf{x}) = \mathbf{p}'(\mathbf{x})\mathbf{A}^{-1}(x)\mathbf{B}(\mathbf{x}), \tag{5.16}$$

$$\boldsymbol{\Phi}_{,i} = \mathbf{p}'_{,i}\mathbf{A}^{-1}\mathbf{B} + \mathbf{p}'\mathbf{A}^{-1}_{,i}\mathbf{B} + \mathbf{p}'\mathbf{A}^{-1}\mathbf{B}_{,i}, \tag{5.17}$$

where $\mathbf{A}^{-1}_{,i} = -\mathbf{A}^{-1}\mathbf{A}_{,i}\mathbf{A}^{-1}$.

It should be noted that the approximation in Eq. (5.15) is no longer a polynomial. However, if $u(\mathbf{x})$ is a polynomial, it will be reproduced exactly by $\hat{u}(\mathbf{x})$. If the weight function $W_I(\mathbf{x})$ and its first kth derivatives are continuous, then the shape function $\mathbf{\Phi}(\mathbf{x})$ and its first kth derivatives will be continuous.

The standard least-squares interpolant is obtained if the weight function is chosen to be constant over the entire domain. However, all the unknowns are then fully coupled. By choosing the weight function to have a large domain of influence, the approximation behaves like a polynomial of higher order than $\mathbf{p}(\mathbf{x})$. Limiting the weight function to be nonzero over a small subdomain results in a sparse system of equations.

The standard finite element formulation will be obtained if the weight function is chosen to be piecewise constant over each subdomain or element.

Choice of Weight Function

The weight functions $W_I(\mathbf{x})$ play an important rule in the performance of the meshless methods. They should be constructed so that they are positive and that a unique solution $\mathbf{a}(\mathbf{x})$ is guaranteed. They are monotonic decreasing functions with respect to the distance from $\mathbf{x}$ to $\boldsymbol{x}_I$.

The commonly used weight functions are functions of the distance between two points, i.e.,

$$W_I(\mathbf{x}) = W(\|\mathbf{x} - \mathbf{x}_I\|). \tag{5.18}$$

We have weight functions of the exponential, cubic spline, and quartic spline, i.e.,

Exponential:

$$W(s_I) = \begin{cases} e^{-(s_I/\alpha)2}, & \text{for } s_I \le 1 \\ 0, & \text{for } s_I > 1 \end{cases} \tag{5.19}$$

or

$$W(s_I) = \begin{cases} \dfrac{e^{-(s_I/c)2} - e^{-(\rho_I/c)2}}{1 - e^{-(\rho_I/c)^2}}, & \text{for } s_I \le \rho_I \\ 0, & \text{for } s_I > \rho_I \end{cases} \tag{5.20}$$

Cubic spline:

$$W(s_I) = \begin{cases} \dfrac{2}{3} - 4s_I^2 + 4s_I^3, & \text{for } s_I \le \dfrac{1}{2} \\ \dfrac{4}{3} - 4s_I + 4s_I^2 - \dfrac{4}{3}s_I^3, & \text{for } \dfrac{1}{2} < s_I \le 1 \\ 0, & \text{for } s_I > 1 \end{cases} \tag{5.21}$$

Quartic spline:

$$W(s_I) = \begin{cases} 1 - 6s_I^2 + 8s_I^3 - 3s_I^4, & \text{for } s_I \le 1 \\ 0, & \text{for } s_I > 1 \end{cases} \tag{5.22}$$

Smooth particle hydrodynamics (SPH) uses a spline weight function:

$$W(s_I) = \frac{2}{3h}\begin{cases} 1-\frac{3}{2}s_I^2+\frac{3}{4}s_I^3, & \text{for } s_I \le 1 \\ \frac{1}{4}[2-s_I]^3, & \text{for } 1 \le s_I \le 2 \\ 0, & \text{for } s_I > 2 \end{cases} \tag{5.23}$$

where α and c are numerical parameters that can be used to adjust the weights and

$$s_I \equiv \frac{R_I}{\rho_I}, \quad R_I \equiv \|\boldsymbol{r}_I\|, \quad \mathbf{r}_I \equiv \mathbf{x} - \mathbf{x}_I. \tag{5.24}$$

It is noticed that $W(s_I) \to \delta(s_I)$ as $\rho_I \to 0$, ρ_I is the radius of the support of the Ith node.

The choice of weight function is more or less arbitrary as long as the weight function is positive and continuous together with its derivatives up to the desired order. The exponential form in Eq. (5.19) is not zero at $s_I = 1$, the parameter $\alpha = 0.4$ results in $W(1) \cong 0.002$, but it performs well in SPH. And it is called the "golden rule of SPH" (Monaghan, 1992). Generally, the exponential functions are computationally more demanding, but they may be less sensitive to the size of the support. The common choice in element-free Galerkin methods (EFG) is the quartic spline, i.e.,

$$W(s_I) = \begin{cases} 1-6s_I^2+8s_I^3-3s_I^4, & \text{for } s_I \le 1 \\ 0, & \text{for } s_I > 1 \end{cases} \tag{5.25}$$

Its derivative is

$$\frac{\partial W_I(\boldsymbol{x})}{\partial \boldsymbol{x}} = \frac{dW(s_I)}{ds_I}\frac{\partial s_I}{\partial \mathbf{x}} = \begin{cases} (-12s_I+24s_I^2-12s_I^3)\mathbf{r}_I/\rho_I R_I, & \text{if } s_I \le 1 \\ 0, & \text{if } s_I > 1 \end{cases} \tag{5.26}$$

It satisfies

$$\begin{gathered} W(0) = 1, \quad W(1) = 0 \\ \left.\frac{dW}{ds_I}\right|_{s_I=0} = 0, \quad \left.\frac{dW}{ds_I}\right|_{s_I=1} = 0, \quad \left.\frac{d^2W}{ds_I^2}\right|_{s_I=1} = 0 \end{gathered} \tag{5.27}$$

These conditions provide continuous first- and second-order derivatives to the weight function and the moving least-square approximations.

The support, or the domain of influence, of the weight function associated with node I is selected to satisfy the following conditions:

1. The support, as reflected by the radius, ρ_I, should be large enough to provide a sufficient number of neighbors at every sample point (integration point) to ensure the invertability of matrix $\mathbf{A} \equiv \sum_{I=1}^{n} W_I(\boldsymbol{x})\mathbf{p}(\mathbf{x}_I)\mathbf{p}'(\mathbf{x}_I)$.
2. The support should be large enough to ensure that information is passed into four quadrants at every sample point, except points near the boundary, so that a given interior sample point has neighbors on all sides.

3. The support should be small enough to provide adequate local character to the least-square approximation.
4. The support should be not too large for the sake of computational cost.

Formulation of Meshless Analysis

The strong form of continuum mechanics is expressed as

$$t_{ij,i} + \rho f_j - \rho \ddot{u}_j = 0 \quad \text{in } \Omega, \tag{5.28}$$

$$u_i = \bar{u}_i \quad \text{on } \Gamma_{ui}, \Gamma_{ui} \tag{5.29}$$

$$t_i \equiv t_{ki} n_k = \bar{t}_i \quad \text{on } \Gamma_{ti}, \Gamma_{ti} \tag{5.30}$$

where the union of the essential boundary, Γ_{ui}, and the natural boundary, Γ_{ti}, is the enclosing surface of the domain Ω, i.e., $\Gamma_{ui} \bigcup \Gamma_{ti} = \partial\Omega$. If there is no ambiguity, we use $\Gamma_u \equiv \bigcup_i \Gamma_{ui}$ and $\Gamma_t \equiv \bigcup_i \Gamma_{ti}$ for abbreviation.

The constrained variational principle yields

$$\int_\Omega t_{ij} \delta e_{ij} \, \mathrm{d}V + \int_\Omega \rho \ddot{u}_i \delta u_i \, \mathrm{d}V - \int_{\Gamma_t} \bar{t}_i \delta u_i \, \mathrm{d}S - \int_\Omega \rho f_i \delta u_i \, \mathrm{d}V$$
$$+ \int_{\Gamma_u} \lambda_i \delta u_i \, \mathrm{d}S + \int_{\Gamma_u} \delta\lambda_i (u_i - \bar{u}_i) \, \mathrm{d}S = 0, \tag{5.31}$$

where λ is the vector of Lagrange multiplies introduced here to enforce the essential boundary conditions, Eq. (5.29), on Γ_u.

Let the approximated displacement field, $\hat{u}_i(\mathbf{x})$, be expressed as

$$u_i \cong \hat{u}_i = \Phi_{i\alpha} U_\alpha. \tag{5.32}$$

Then the strain field, $\mathbf{e}(\mathbf{x})$, can be obtained as

$$e_{ij} \cong \hat{e}_{ij} = \frac{1}{2}(\Phi_{i\alpha,j} + \Phi_{j\alpha,i}) U_\alpha = B_{ij\alpha} U_\alpha. \tag{5.33}$$

The detailed expressions for the shape function, $\mathbf{\Phi}$, and its derivative, $\Phi_{,i}$, for moving least-square approximation have been obtained previously. For other kinds of approximations, the expressions will be different, but the procedure remains the same.

Also, approximate λ on Γ_u in terms of nodal value Λ as

$$\lambda_i = \psi_{i\beta} \Lambda_\beta, \quad \beta = 1, 2, 3, \ldots, l, \tag{5.34}$$

where l is the number of nodes whose weight functions are nonzero on the essential boundary. We then have

$$\delta\lambda_i = \psi_{i\beta} \delta\Lambda_\beta, \tag{5.35}$$

$$\delta u_i = \mathbf{\Phi}_{i\alpha} \delta U_\alpha, \tag{5.36}$$

$$\delta e_{ij} = \frac{1}{2}(\mathbf{\Phi}_{i\alpha,j} + \mathbf{\Phi}_{j\alpha,i}) \delta U_\alpha = B_{ij\alpha} \delta U_\alpha. \tag{5.37}$$

One of the differences between the finite element methods and the meshless methods is that Eq. (5.15) is an approximation rather than an interpolation, i.e.,

$$\mathbf{\Phi}(\mathbf{x}_I)\mathbf{U} \neq U_I, \tag{5.38}$$

and therefore the essential boundary conditions, Eq. (5.29), should be read as

$$\Phi_{i\alpha} U_\alpha = \bar{u}_i \quad \text{on } \Gamma_u, \tag{5.39}$$

which are the constraints rather than the specifications on $\mathbf{U}$.

For viscoelastic solid, the constitutive relation is of the form

$$t_{ij} = A_{ijkl} e_{kl} + a_{ijkl} \dot{e}_{kl}. \tag{5.40}$$

Equation (5.31) then becomes

$$\begin{aligned}\delta\bar{\Pi} = &\int_\Omega \{A_{ijkl} B_{kl\beta} U_\beta B_{ij\alpha} \delta U_\alpha + a_{ijkl} B_{kl\beta} \dot{U}_\beta B_{ij\alpha} \delta U_\alpha + \rho \Phi_{i\beta} \ddot{U}_\beta \Phi_{i\alpha} \delta U_\alpha)\,\mathrm{d}V \\ &- \int_{\Gamma_t} \bar{t}_i \Phi_{i\alpha} \delta U_\alpha \,\mathrm{d}S - \int_\Omega \rho f_i \Phi_{i\alpha} \delta U_\alpha \,\mathrm{d}V + \int_{\Gamma_u} \psi_{i\gamma} \Lambda_\gamma \delta U_\alpha \,\mathrm{d}S \\ &+ \int_{\Gamma_u} \psi_{i\gamma} \delta\Lambda_\gamma (\Phi_{i\alpha} U_\alpha - \bar{u}_i)\,\mathrm{d}S = 0. \end{aligned} \tag{5.41}$$

This can be rewritten as

$$\delta U_\alpha \{M_{\alpha\beta} \ddot{U}_\beta + C_{\alpha\beta} \dot{U}_\beta + K_{\alpha\beta} U_\beta - F_\alpha + G_{\alpha\gamma} \Lambda_\gamma\} + \delta\Lambda_\gamma \{G_{\alpha\gamma} U_\alpha - f_\gamma\} = 0, \tag{5.42}$$

where

$$M_{\alpha\beta} = \int_\Omega \rho \Phi_{i\beta} \Phi_{i\alpha} \,\mathrm{d}\Omega = M_{\beta\alpha}, \tag{5.43}$$

$$C_{\alpha\beta} = \int_\Omega a_{ijkl}(\boldsymbol{x}) B_{kl\beta} B_{ij\alpha}(\boldsymbol{x}) \,\mathrm{d}\Omega(\boldsymbol{x}), \tag{5.44}$$

$$K_{\alpha\beta} = \int_\Omega A_{ijkl} B_{ij\alpha}(\boldsymbol{x}) B_{kl\beta}(\boldsymbol{x}) \,\mathrm{d}\Omega(\boldsymbol{x}), \tag{5.45}$$

$$G_{\alpha\gamma} = \int_{\Gamma_u} \Phi_{i\alpha} \psi_{i\gamma} \,\mathrm{d}s, \tag{5.46}$$

$$F_\alpha = \int_\Omega \rho f_j \Phi_{j\alpha} \,\mathrm{d}\Omega + \int_{\Gamma_t} \bar{t}_j \Phi_{j\alpha} \,\mathrm{d}S, \tag{5.47}$$

$$f_\gamma = \int_{\Gamma_u} \psi_{i\gamma} \bar{u}_i \,\mathrm{d}S. \tag{5.48}$$

Equation (5.42) should hold for any arbitrary $\delta\mathbf{U}$ and $\delta\Lambda$. The governing equations in matrix form are thus obtained:

$$\mathbf{M}\ddot{\mathbf{U}} + \mathbf{C}\dot{\mathbf{U}} + \mathbf{K}\mathbf{U} + \mathbf{G}\boldsymbol{\Lambda} = \mathbf{F}, \tag{5.49}$$

$$\mathbf{G}'\mathbf{U} = \mathbf{f}. \tag{5.50}$$

For static problems, the governing equations are reduced to

$$\begin{vmatrix} \mathbf{K} & \mathbf{G} \\ \mathbf{G}^t & 0 \end{vmatrix} \begin{vmatrix} \mathbf{U} \\ \mathbf{\Lambda} \end{vmatrix} = \begin{vmatrix} \mathbf{F} \\ \mathbf{f} \end{vmatrix}. \tag{5.51}$$

For dynamic problems, one needs to solve the following system of linear equations to get $\bar{\mathbf{K}}'$ and $\bar{\mathbf{G}}'$ first (Chen et al., 2002)

$$\begin{vmatrix} \mathbf{K}^t & \mathbf{G} \\ \mathbf{G}^t & \mathbf{0} \end{vmatrix} \begin{vmatrix} \bar{\mathbf{K}}^t \\ \bar{\mathbf{G}}^t \end{vmatrix} = \begin{vmatrix} \mathbf{I} \\ \mathbf{0} \end{vmatrix}. \tag{5.52}$$

Then the governing equations for the displacements become

$$\bar{\mathbf{K}}\mathbf{M}\ddot{\mathbf{U}} + \bar{\mathbf{K}}\mathbf{C}\dot{\mathbf{U}} + \mathbf{U} = \bar{\mathbf{K}}\mathbf{F} + \bar{\mathbf{G}}\mathbf{f}. \tag{5.53}$$

It can be verified both analytically and numerically that the displacement field so obtained satisfies Eq. (5.29), the essential boundary conditions.

Evaluation of the Integral

The major dilemma in meshless methods revolves around how to evaluate the integrals in the weak form, in other words, how to obtain the matrices in Eqs. (5.43)–(5.48). Several approaches have been proposed and studied and can be summarized as follows.

Nodal Integration

Partial integration was achieved by evaluating the integrals of the weak form only at the nodes (Beissel and Belytschko, 1996). The integration is evaluated by

$$\int_{\Omega} f(\mathbf{x})\,\mathrm{d}V = \sum_{I=1}^{N} f(\mathbf{x}_I)\Delta V_I. \tag{5.54}$$

Therefore, no background mesh is needed. It is a very fast approach. However, it would result in a spatial instability, and a stabilization procedure is then needed. It was suggested by Belytschko et al. (2000) that the best approach to stabilize particle discretization of solids and fluids is to use Lagrangian kernels with stress points.

Background Mesh

A regular array of domain in the background is used for quadrature as shown in Fig. 5.1. It is also called cell or octree quadrature method.

Another way is to use a finite element mesh, as shown in Fig. 5.2, as the background mesh. The procedure is then similar to that in finite element method, i.e., the integral is performed by Gauss quadrature.

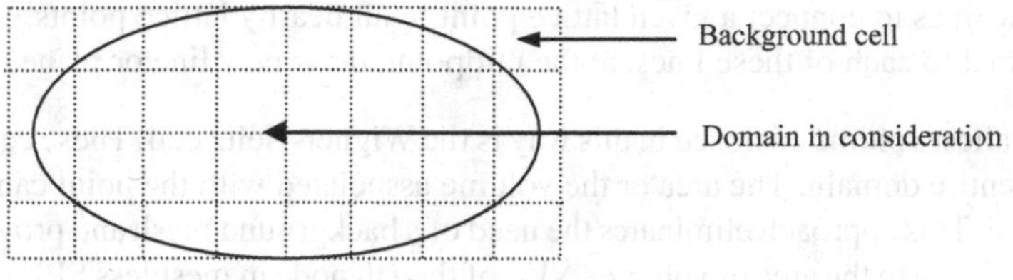

FIGURE 5.1. Cell quadrature.

Wigner–Seitz Cell

This approach is adopted in lattice dynamics. In lattice dynamics, the cell associated with the lattice point can be obtained by forming a *Wigner–Seitz primitive cell.* The procedures to obtain a Wigner–Seitz cell are (cf. Fig. 5.3):

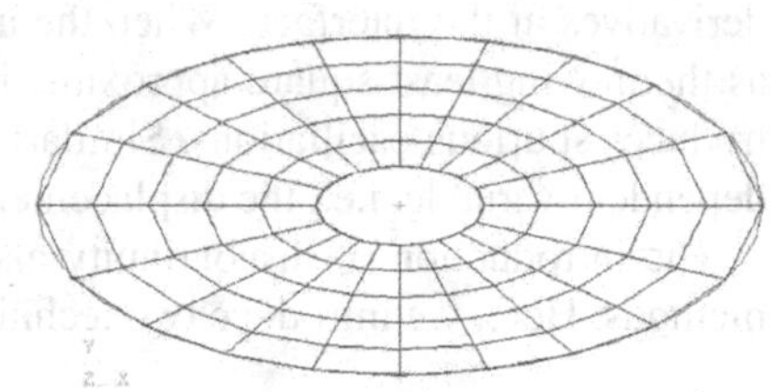

FIGURE 5.2. Finite element mesh.

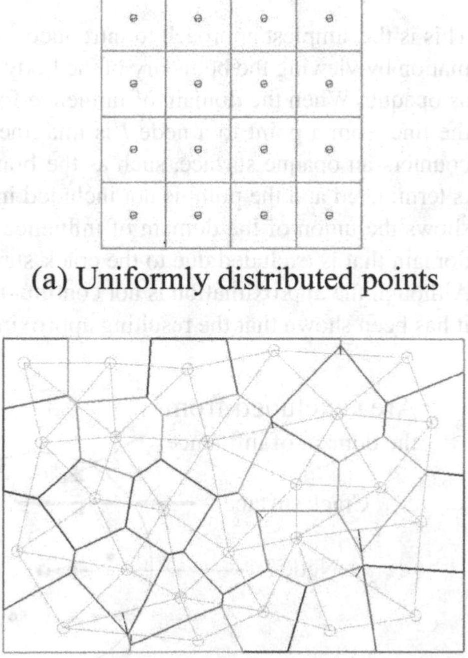

FIGURE 5.3. Wigner–Seitz cell: (a) uniformly distributed points, (b) nonuniformly distributed points.

1. Draw lines to connect a given lattice point to all nearby lattice points.
2. Normal to each of these lines, at the midpoint, draw new line or plane.

The smallest volume enclosed in this way is the Wigner–Seitz cell. These cells then fill the entire domain. The area or the volume associated with the point can be determined. This approach eliminates the need of a background mesh and provides an exact solution to the area or volume, ΔV_I, of the Ith node in meshless SPH or EFG.

Treatment of Discontinuity

Many engineering problems involve multiconnected domains and various kinds of discontinuities. For example, in elastostatics, for problems involving two materials, the coefficients in the partial differential equations are discontinuous across the interface between the materials. This results in solutions with discontinuous derivatives at the interface. When the approximation is a smooth function, such as the moving least-square approximation, the discontinuity in the derivative introduces spurious oscillations. Similarly, when a crack is modeled in a body, the dependent variable, i.e., the displacement must be discontinuous across the crack.

The introduction of discontinuity also requires special treatment in meshless methods. Here, we introduce two techniques (cf. Belytschko et al., 1996).

Visibility Criterion

This is the simplest approach to introduce a discontinuity into meshless approximation by viewing the boundary of the body and any interior line of discontinuity as opaque. When the domain of influence for the weight function is constructed, the line from a point to a node I is imagined to be a ray of light. If the ray encounters an opaque surface, such as the boundary of body or a crack surface, it is terminated and the point is not included in the domain of influence. Figure 5.4 shows the union of the domain of influence of node I near the crack tip and the domain that is excluded due to the crack surface and the indicated "ray of light." Although the approximation is not continuous across these lines of discontinuity, it has been shown that the resulting approximations still leads to convergence.

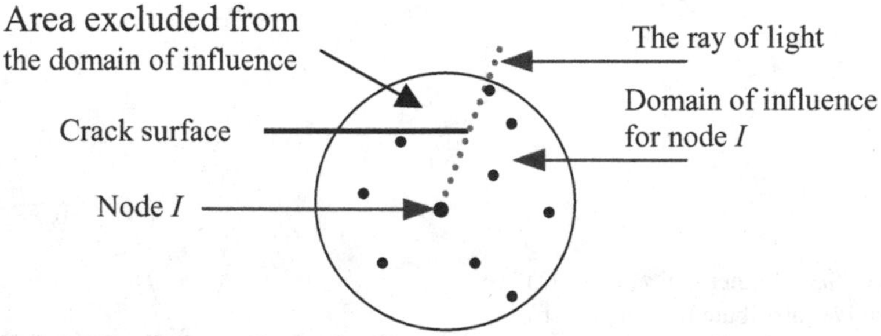

FIGURE 5.4. Domain of influence of node I adjacent to a crack tip by visibility criterion.

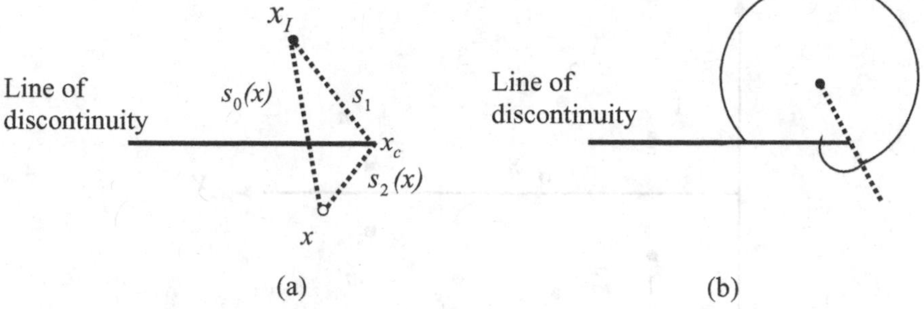

FIGURE 5.5. Domain of influence of node I near the tip of discontinuity by diffraction method: (a) scheme for the diffraction method, (b) domain of influence of the weight function near the tip of discontinuity.

Diffraction Method

This method is motivated by the way light diffracts around a sharp corner. The essence of the diffraction method is to treat the line of discontinuity as opaque but to evaluate the length of the ray s by a path that passes around the corner of the discontinuity, as shown in Fig. 5.5a. The weight parameter s is computed by

$$s(\boldsymbol{x}) = \left(\frac{s_1 + s_2(\boldsymbol{x})}{s_0(\boldsymbol{x})}\right)^{\lambda} s_0(\boldsymbol{x}), \tag{5.55}$$

where

$$s_0(\boldsymbol{x}) = \|\boldsymbol{x} - \boldsymbol{x}_I\|, \quad s_1 = \|\boldsymbol{x}_C - \boldsymbol{x}_\mathbf{I}\|, \quad s_2(\boldsymbol{x}) = \|\boldsymbol{x} - \boldsymbol{x}_C\|. \tag{5.56}$$

As a consequence, the domain of influence is determined by weight function $W(s_I)$, which, as can be seen in Fig. 5.5b, includes the point on the other side of the line of discontinuity. The weight function and shape function are continuous within the domain but are discontinuous across the line of discontinuity.

Treatment of Mirror Symmetry

The treatment of mirror symmetry in meshless method is very different from that in finite element method. It needs special and careful treatment. It will lead to erroneous results otherwise. To make this point, a specimen having mirror symmetry with respect to the x–z plane is illustrated in Fig. 5.6. It is seen that there is a sampling point above the mirror plane, marked by its coordinate $\boldsymbol{x}$, surrounded by many supportive nodes, among which those below the mirror plane are labeled with $\{1', 2', \ldots, m'\}$, and correspondingly, their mirror images are labeled with $\{1, 2, \ldots, m\}$. The displacements of the sampling point can now be expressed as

$$u_i(\mathbf{x}) = \sum_{I=1}^{n} \mathbf{\Phi}(\mathbf{x}) u_i(I), \tag{5.57}$$

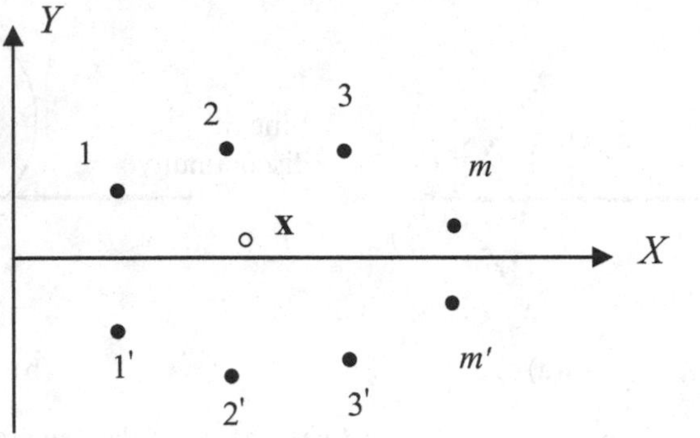

FIGURE 5.6. Illustration of mirror symmetry.

where $u_i(I)$ is the nodal value of u_i at the Ith node. Because of mirror symmetry, we have

$$\rho_{J'} = \rho_J, \quad x_{J'} = x_J, \quad z_{J'} = z_J, \quad y_{J'} = -y_J, \tag{5.58}$$

$$u_x(J') = u_x(J), \quad u_z(J') = u_z(J), \quad u_y(J') = -u_y(J), \tag{5.59}$$

where $J = 1, 2, \ldots, m$. Now, Eq. (5.57) can be rewritten as

$$u_x(\mathbf{x}) = \sum_{J=1}^{m} \{\Phi_J + \Phi_{J'}\} u_x(J) + \text{other terms}, \tag{5.60}$$

$$u_y(\mathbf{x}) = \sum_{J=1}^{m} \{\Phi_J - \Phi_{J'}\} u_y(J) + \text{other terms}, \tag{5.61}$$

$$u_z(\mathbf{x}) = \sum_{J=1}^{m} \{\Phi_J + \Phi_{J'}\} u_z(J) + \text{other terms}, \tag{5.62}$$

where "other terms" means the contribution from other supportive nodes above the mirror plane whose mirror images do not support the sampling point $\mathbf{x}$. On the contrary, the mirror symmetry with respect to x–z plane in finite element method is treated by simply setting $u_y(\mathbf{x}) = 0$ for all nodes on the x–z plane.

H- and P-Refinements

The essential characteristics of the meshless methods are that there is no need for a highly structured mesh as required in the finite element methods. Obviously, the major advantages of meshless methods must be closely related to those characteristics.

In finite element methods, to enhance the accuracy in a critical region, e.g., around the crack tip, one can use the techniques of h-refinement and/or p-refinement. The h-refinement calls for a finer mesh in that critical region. The p-refinement needs a higher order polynomial, which is corresponding to higher

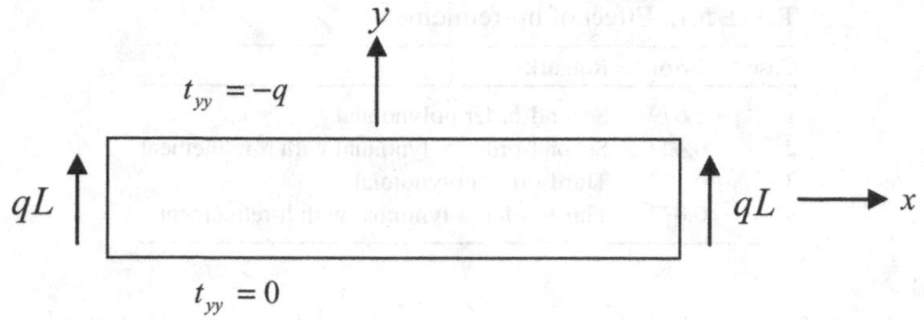

FIGURE 5.7. Simply supported beam subject to uniform loading at the top surface.

order elements, e.g., the four-node element has to be upgraded to eight-node element in a plane problem. In either case, the finite element mesh has to be regenerated with tremendous cost.

On the other hand, the h-refinement in meshless methods can be achieved by simply sprinkling arbitrary number of nodes in the critical region. This means that only the number and the locations of the added nodes need to be specified. Even simpler for the p-refinement, only the order of polynomial needs to be modified. The following is an example to show the improvement of the accuracy upon h- and p-refinements.

Consider an elastic problem of a beam, with length $2L$ (from $x = -L$ to $x = L$) and height $2c$ (from $y = -c$ to $y = c$), simply supported at the two ends and subjected to a uniformly distributed load $t_{yy} = -q$ at the top surface ($y = c$) as shown in Fig. 5.7. The analytical elastic solution of the beam in the case of plane stress can be expressed as:

$$t_{xx} = -0.5q[(L^2 - x^2)y + 2y^3/3 - 0.4c^2y]/I, \quad (5.63)$$

$$t_{yy} = 0.5q[y^3/3 - c^2y - 2c^3/3]/I, \quad (5.64)$$

$$t_{xy} = -0.5q(y^2 - c^2)x/I, \quad (5.65)$$

where $I = 2c^3/3$ is the moment of inertia of the beam. This solution indicates the following boundary conditions:

$$\text{At } y = c, \quad t_{xy} = 0, \quad t_{yy} = -q. \quad (5.66)$$

$$\text{At } y = -c, \quad t_{xy} = t_{yy} = 0. \quad (5.67)$$

$$\text{At } x = L \text{ and } x = -L,$$

$$F_x = \int_{y=-c}^{y=c} t_{xx}\,\mathrm{d}y = 0, \quad (5.68)$$

$$F_y(x = L) \equiv \int_{y=-c}^{y=c} t_{yx}\mathrm{d}y = qL$$

$$F_y(x = -L) \equiv \int_{y=-c}^{y=c} -t_{yx}\mathrm{d}y = qL \quad (5.69)$$

TABLE 5.1. Effect of hp-refinements.

Case	Error	Remarks
1	2.079	Second-order polynomial
2	0.427	Second-order polynomial with h-refinement
3	0.152	Third-order polynomial
4	0.077	Third-order polynomial with h-refinement

Define the differences between the exact analytical stresses and the meshless stresses as

$$\Delta t_{ij} = t_{ij}(\text{meshless}) - t_{ij}(\text{analytical}), \quad (i, j = 1, 2); \tag{5.70}$$

then the error, E, is defined as the dimensionless Euclidean norm of the differences at the sampling points as follows:

$$E \equiv \left[\sum_{i=1}^{m} (\Delta t_{xx}^2 + \Delta t_{yy}^2 + \Delta t_{xy}^2)_i \right]^{1/2} /q/3m, \tag{5.71}$$

where m is the number of sampling points in question. For this typical beam problem, the errors for four cases are listed in Table 5.1 in which the effectiveness of hp-refinements is demonstrated.

Meshless methods possess a favorable environment for h- and p-refinements. Based on our working experience with the meshless method, we now summarize the general characteristics of four refinement approaches as follows:

1. *P-refinement: raise the order of polynomials*. In this approach, there is no need to increase the number of nodes. However, in many situations, we need to enlarge the size of support of certain nodes to ensure the invertability of the **A** matrix (cf. Eq. (5.14)).

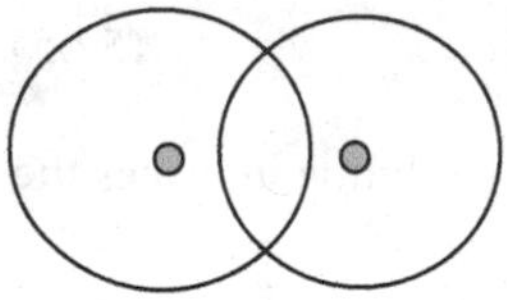

FIGURE 5.8. P-refinement.

2. *H-refinement: simply add nodes*. In this approach we simply add nodes, but the size of the support (radius of influence) of each existing and newly added node is not changed. This method can improve the accuracy, but the effect may become saturated after too many nodes are added.

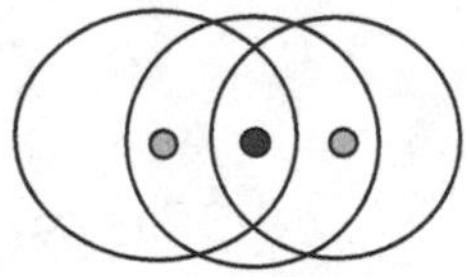

FIGURE 5.9. H-refinement without change of the size of support.

3. *H-refinement: add more nodes and also decrease the size of support.* In this approach, the size of support of all nodes is reduced. It would work to improve the accuracy in the region where nodes are added. Because the increase of the number of nodes is proportional to the decrease of the size of support, this approach may be costly.

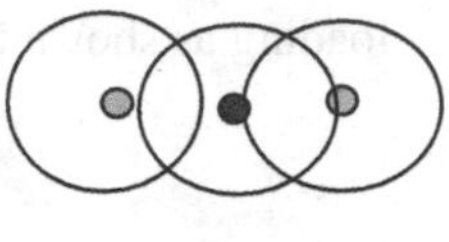

FIGURE 5.10. H-refinement with the reduction of size of support.

4. *H-refinements: add nodes and the newly added nodes have support with smaller size.* In this approach, the size of support of those newly added nodes is smaller than that of the previously existing nodes. It surely works to improve the accuracy in the region where nodes are added. It seems to be the best way of h-refinement to improve the accuracy with a minimal cost.

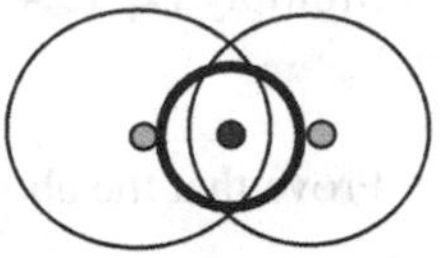

FIGURE 5.11. H-refinement with change of the size of support of newly added nodes.

In Figs. 5.8–5.11, the red dot is a newly added node. It would be worthwhile to perform a systematic study on the accuracy and cost of a certain benchmark problem that has exact and analytical solution using these four approaches.

This chapter has presented general procedures of meshless methods, including the construction of the approximation, choices of weight function, formulation of meshless analysis, evaluation of the integrals, treatment of discontinuity, treatment of mirror symmetry, and H- and P-refinements. Note that meshless method is still in its early stage of development. Further advances of these procedures and many other aspects of meshless methods will deepen our understanding of meshless methods. A collaboration between engineers and mathematicians can be of great benefit.

Problems

1. For two-dimensional problems with second-order and third-order polynomial bases, at least how many sampling points do you need to ensure the invertability of matrix **A**?
2. What are the differences between finite element method and meshless method in the treatment of boundary conditions, discontinuity, refinements, and symmetry? Why?
3. What is the major cost in meshless computation? Why?

4. Recall that the constitutive equation for viscoelastic solid as

$$t_{ij} = A_{ijkl}e_{kl} + a_{ijkl}\dot{e}_{kl}$$

If $A_{ijkl} = A_{klij}$ and $a_{ijkl} = a_{klij}$ (cf. Chapter 2), show that the stiffness matrix **K**, Eq. (5.45), and the damping matrix **C**, Eq. (5.44), are symmetric.

5. Prove that the stresses, Eqs. (5.63)–(5.65), of the beam subjected to uniform loading as shown in Fig. 5.7 satisfy the equilibrium equations

$$\frac{\partial t_{xx}}{\partial x} + \frac{\partial t_{yx}}{\partial y} = 0$$

$$\frac{\partial t_{xy}}{\partial x} + \frac{\partial t_{yy}}{\partial y} = 0$$

6. From Eq. (5.52), it is seen that

$$\mathbf{G}'\bar{\mathbf{K}}' = 0 \Rightarrow \bar{\mathbf{K}}\mathbf{G} = 0$$
$$\mathbf{K}'\bar{\mathbf{K}}' + \mathbf{G}\bar{\mathbf{G}}' = \mathbf{I} \Rightarrow \bar{\mathbf{K}}\mathbf{K} + \bar{\mathbf{G}}\mathbf{G}' = \mathbf{I}$$

Multiply Eq. (5.49) by $\bar{\mathbf{K}}$, it results

$$\bar{\mathbf{K}}\{\mathbf{M}\ddot{\mathbf{U}} + \mathbf{C}\dot{\mathbf{U}} + \mathbf{K}\mathbf{U} + \mathbf{G}\mathbf{\Lambda}\} = \bar{\mathbf{K}}\mathbf{F}$$

Prove that the above equation leads to

$$\bar{\mathbf{K}}\mathbf{M}\ddot{\mathbf{U}} + \bar{\mathbf{K}}\mathbf{C}\dot{\mathbf{U}} + \mathbf{U} = \bar{\mathbf{K}}\mathbf{F} + \bar{\mathbf{G}}\mathbf{f}$$

Also prove that the essential boundary conditions are identically satisfied.

6 Meshless Analysis of Elastic Problems

The mechanical properties of materials include elastic, anelastic, or inelastic behavior. A solid is said to be *elastic* when any deformations disappear "quickly" once the external forces are removed (i.e., the solid returns to its initial state). In the elastic region, the deformations are directly proportional to the external forces, i.e., the mechanical strain tensor is a linear function of stress tensor. The *anelastic* behavior of materials is a type of time-dependent mechanical behavior in which the applied stresses and the resulting strain are not uniquely related to each other due to relaxation effects. Other types of response, including permanent deformations, plasticity, viscoelasticity, etc., correspond to *inelastic* behavior.

In this chapter, we consider the meshless analysis of elastic problems, for both static and dynamic cases. A detailed analysis of crack propagation problem is also included. A meshless computer program for elastostatics and elastodynamics is posted on this book's page at www.springeronline.com.

Background Theory for Applications of Elastostatics

Plane Stress Problem

Consider some typical plane stress problems shown in Fig. 6.1. Typically, a thin plate is subjected to loads applied in the xy plane that is the plane of the structure. The thickness of the plate is small compared with the dimensions in the xy plane, so the stresses are assumed to be constant through the thickness of the plane and t_{zz}, t_{zx}, and t_{zy} are ignored. Thus, the displacements are expressed as

$$\mathbf{u} = \{u, v\}^t, \tag{6.1}$$

where u and v are the in-plane displacements in the x and y directions, respectively.

The strain components expressed in vector notation as

$$\varepsilon = [\varepsilon_x, \varepsilon_y, \gamma_{xy}]^t, \tag{6.2}$$

where, for linear small strain theory, the normal strains are given as

$$\varepsilon_x = \frac{\partial u}{\partial x} \quad \text{and} \quad \varepsilon_y = \frac{\partial v}{\partial y}, \tag{6.3}$$

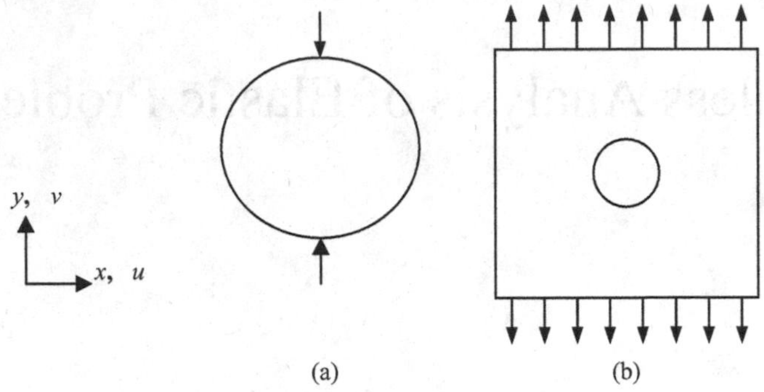

FIGURE 6.1. Plane stress problem. (a) Circular disc under point loads. (b) Thin plate with a hole under tension.

and the shear strain is given as

$$\gamma_{xy} = \frac{\partial u}{\partial y} + \frac{\partial v}{\partial x}. \tag{6.4}$$

The linear stress–strain relationships can be written as

$$\boldsymbol{\sigma} = \boldsymbol{D}\boldsymbol{\varepsilon}, \tag{6.5}$$

where

$$\boldsymbol{\sigma} \equiv [t_{xx}, t_{yy}, t_{xy}]^t, \tag{6.6}$$

and for isotropic material

$$\boldsymbol{D} = \frac{E}{(1-\upsilon^2)} \begin{bmatrix} 1 & \upsilon & 0 \\ \upsilon & 1 & 0 \\ 0 & 0 & (1-\upsilon)/2 \end{bmatrix}. \tag{6.7}$$

E and υ are the Young's modulus and Poisson's ratio, respectively.

The body forces **f** can be written as

$$\mathbf{f} = \{f_x, f_y\}^t, \tag{6.8}$$

in which f_x and f_y are the body forces per unit mass in the x and y directions, respectively.

The nonessential boundary conditions may be expressed in terms of surface tractions $\bar{\mathbf{t}}$:

$$\bar{\mathbf{t}} = \{\bar{t}_x, \bar{t}_y\}^t, \tag{6.9}$$

in which $\bar{t}_x$ and $\bar{t}_y$ are the surface tractions per unit length.

Plane Strain Problem

For plane strain problems, the thickness dimension normal to the xy plane is large compared with the typical dimensions in the xy plane and the body is subjected

FIGURE 6.2. Plane strain problem: long cylinder under internal pressure.

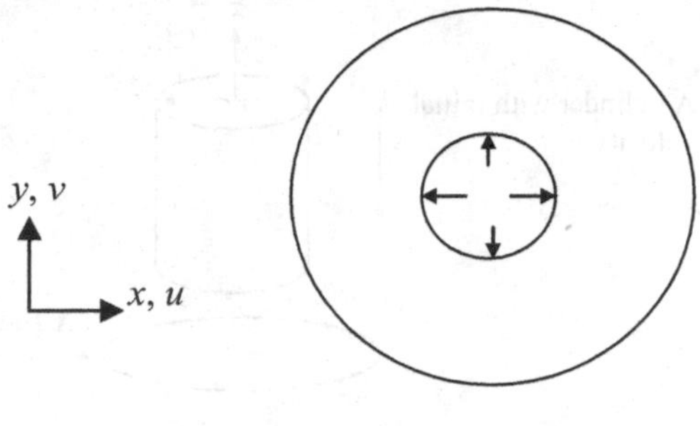

to loads in the xy plane only. It may be assumed that the strains in the z direction are negligible and that the in-plane displacements u and v are independent of z. A typical plane strain problem is illustrated in Fig. 6.2.

We then have the displacement vector as

$$\mathbf{u} = \{u, v\}^t, \tag{6.10}$$

the in-plane strain

$$\boldsymbol{\varepsilon} = [\varepsilon_x, \varepsilon_y, \gamma_{xy}]^t, \tag{6.11}$$

and the in-plane stress

$$\boldsymbol{\sigma} \equiv [t_{xx}, t_{yy}, t_{xy}]^t, \tag{6.12}$$

in which u and v are the in-plane displacements in the x and y directions, respectively. The stress–strain relationships have the same form as that in the plane stress problem, i.e.,

$$\boldsymbol{\sigma} = \boldsymbol{D}\boldsymbol{\varepsilon}. \tag{6.13}$$

For linear isotropic elastic materials, we have $\boldsymbol{D}$ as

$$\boldsymbol{D} = \frac{E}{(1+\upsilon)(1-2\upsilon)} \begin{bmatrix} (1-\upsilon) & \upsilon & 0 \\ \upsilon & (1-\upsilon) & 0 \\ 0 & 0 & (1-2\upsilon)/2 \end{bmatrix}. \tag{6.14}$$

Note that the stress normal to the xy plane is nonzero and may be evaluated as

$$t_{zz} = \upsilon(t_{xx} + t_{yy}). \tag{6.15}$$

The body forces $\mathbf{f}$ and surface tractions $\bar{\mathbf{t}}$ have the same form as those for plane stress problem.

Axisymmetric Solids

For a three-dimensional (3D) solid which is symmetrical about its centerline axis (which coincides with the z axis) and subjected to loads that are symmetrical about this axis, the behavior is independent of the circumferential coordinate θ.

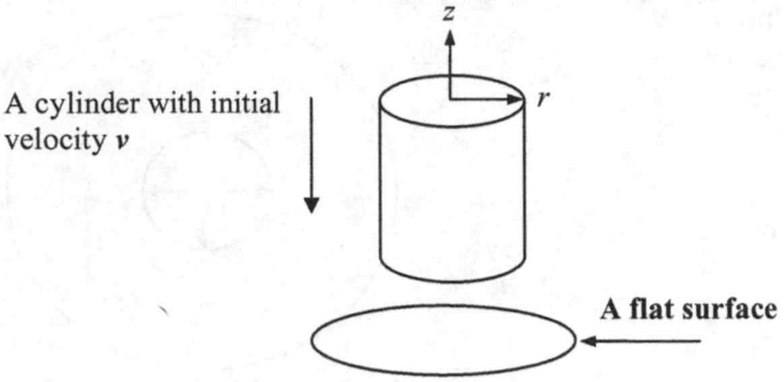

FIGURE 6.3. A typical axisymmetric problem: a cylinder impacts against a fixed flat surface with initial velocity $\boldsymbol{v}$.

A cylinder under impression or impact with a flat surface and (cf. Fig. 6.3) a cylindrical bar under axisymmetric loading are typical axisymmetric problems. For axisymmetric problems, the displacements can be expressed as

$$\mathbf{u} = \{u, w\}^t, \tag{6.16}$$

where u and w are the displacements in the r and z directions, respectively.

The nonzero strains are given as

$$\boldsymbol{\varepsilon} = [\varepsilon_r, \varepsilon_\theta, \varepsilon_z, \gamma_{rz}]^t, \tag{6.17}$$

where, for small displacements,

$$\varepsilon_r = \frac{\partial u}{\partial r}, \tag{6.18}$$

$$\varepsilon_\theta = \frac{u}{r}, \tag{6.19}$$

$$\varepsilon_z = \frac{\partial w}{\partial z}, \tag{6.20}$$

$$\gamma_{rz} = \frac{\partial u}{\partial z} + \frac{\partial w}{\partial r}. \tag{6.21}$$

The stress–strain relationships still has the same form as in two-dimensional (2D) linear elasticity

$$\boldsymbol{\sigma} = \boldsymbol{D}\boldsymbol{\varepsilon}, \tag{6.22}$$

where $\boldsymbol{\sigma} \equiv \{t_{rr}, t_{\theta\theta}, t_{zz}, t_{rz}\}^t$ in which t_{rr}, $t_{\theta\theta}$, and t_{zz} are the normal stresses in the r, θ, and z directions, respectively, and t_{rz} is the shear stress in the rz plane, and for isotropic material,

$$\boldsymbol{D} = \frac{E}{(1+\upsilon)(1-2\upsilon)} \begin{bmatrix} (1-\upsilon) & \upsilon & \upsilon & 0 \\ \upsilon & (1-\upsilon) & \upsilon & 0 \\ \upsilon & \upsilon & (1-\upsilon) & 0 \\ 0 & 0 & 0 & (1-2\upsilon)/2 \end{bmatrix}. \tag{6.23}$$

The body forces are

$$\mathbf{f} = \{f_r, f_z\}^t, \tag{6.24}$$

and the boundary surface tractions are

$$\bar{\mathbf{t}} = \{\bar{t}_r, \bar{t}_z\}^t. \tag{6.25}$$

Three-Dimensional Stress Analysis

The displacements in 3D problem can be expressed as

$$\mathbf{u} = \{u, v, w\}^t, \tag{6.26}$$

where u, v, and w are the displacements in the global x, y, and z directions, respectively.

The strain components are

$$\varepsilon = \{\varepsilon_{xx}, \varepsilon_{yy}, \varepsilon_{zz}, \gamma_{yz}, \gamma_{zx}, \gamma_{xy}\}^t, \tag{6.27}$$

corresponding to the stresses

$$\sigma = \{t_{xx}, t_{yy}, t_{zz}, t_{xy}, t_{rz}, t_{zx}\}^t, \tag{6.28}$$

in which t_{xx}, t_{yy}, and t_{zz} are the normal stresses and t_{xy}, t_{yz}, and t_{zx} are the shear stresses, and for small displacements, the normal strains are given as

$$\varepsilon_{xx} = \frac{\partial u}{\partial x}, \quad \varepsilon_{yy} = \frac{\partial v}{\partial y}, \quad \text{and} \quad \varepsilon_{zz} = \frac{\partial w}{\partial z}, \tag{6.29}$$

and the shear strains are given as

$$r_{xy} = \frac{\partial u}{\partial y} + \frac{\partial v}{\partial x}, \quad r_{yz} = \frac{\partial v}{\partial z} + \frac{\partial w}{\partial y}, \quad \text{and} \quad r_{zx} = \frac{\partial w}{\partial x} + \frac{\partial u}{\partial z}. \tag{6.30}$$

For isotropic material, the stiffness matrix in the stress–strain relationship is given as

$$D = \alpha_1 \begin{bmatrix} 1 & \alpha_2 & \alpha_2 & 0 & 0 & 0 \\ \alpha_2 & 1 & \alpha_2 & 0 & 0 & 0 \\ \alpha_2 & \alpha_2 & 1 & 0 & 0 & 0 \\ 0 & 0 & 0 & \alpha_3 & 0 & 0 \\ 0 & 0 & 0 & 0 & \alpha_3 & 0 \\ 0 & 0 & 0 & 0 & 0 & \alpha_3 \end{bmatrix}, \tag{6.31}$$

where

$$\alpha_1 = \frac{E(1-\upsilon)}{(1+\upsilon)(1-2\upsilon)}, \quad \alpha_2 = \frac{\upsilon}{1-\upsilon}, \quad \text{and} \quad \alpha_3 = \frac{(1-2\upsilon)}{2(1-\upsilon)}. \tag{6.32}$$

The body forces $\mathbf{f}$ are

$$\mathbf{f} = \{f_x, f_y, f_z\}^t, \tag{6.33}$$

and the surface tractions $\bar{\mathbf{t}}$ are

$$\bar{\mathbf{t}} = \{\bar{t}_x, \bar{t}_y, \bar{t}_z\}^t. \tag{6.34}$$

Meshless Analysis of Elastostatic Problems

Meshless Formulation of Elastostatics

The meshless formulation for elastostatic problem has been obtained in Chapter 5 as

$$\begin{vmatrix} \mathbf{K} & \mathbf{G} \\ \mathbf{G}^t & \mathbf{0} \end{vmatrix} \begin{vmatrix} \mathbf{U} \\ \mathbf{\Lambda} \end{vmatrix} = \begin{vmatrix} \mathbf{F} \\ \mathbf{f} \end{vmatrix}, \tag{6.35}$$

where $\mathbf{K}$ is the stiffness matrix, $\mathbf{U}$ is the displacement vector, and $\mathbf{\Lambda}$ is the vector of Lagrangian multipliers, and

$$K_{\alpha\beta} = \int_\Omega A_{ijkl} B_{ij\alpha}(\mathbf{x}) B_{kl\beta}(\mathbf{x})\, \mathrm{d}V, \tag{6.36}$$

$$G_{\alpha\beta} = \int_{\Gamma_u} \Phi_{i\alpha} \psi_{i\beta}\, \mathrm{d}S, \tag{6.37}$$

$$F_\alpha = \int_\Omega \rho f_i \Phi_{i\alpha}\, \mathrm{d}V + \int_{\Gamma_t} \bar{t}_i \Phi_{i\alpha}\, \mathrm{d}S, \tag{6.38}$$

$$f_\alpha = \int_{\Gamma_u} \bar{u}_i \psi_{i\alpha}\, \mathrm{d}S. \tag{6.39}$$

After evaluation of those matrices by numerical integrations, the step left is to solve the linear algebra equations in Eq. (6.35). Essentially, there are two different classes of methods for the solutions of Eq. (6.35): direct solution techniques and iterative solution methods. Direct technique can be applied to almost any set of simultaneous linear equations, while for large systems iterative method can be much more effective.

Gauss Elimination

The most effective direct solution technique currently used is basically the application of Gauss elimination. We begin this subject by an example

$$\begin{aligned} 2u + v + w &= 5, \\ 4u - 6v &= -2, \\ -2u + 7v + 2w &= 9. \end{aligned} \tag{6.40}$$

The method to find the unknown value of u, v, and w by Gauss elimination starts by subtracting multiples of the first equation from the others, so as to eliminate u from the last two equations. This results an equivalent system of equations

$$\begin{aligned} 2u + v + w &= 5, \\ -8v - 2w &= -12, \\ 8v + 3w &= 14. \end{aligned} \tag{6.41}$$

We then subtract the second equation from the third so as to eliminate v from the third equation; this gives

$$\begin{aligned} 2u + v + w &= 5, \\ -8v - 2w &= -12, \\ w &= 2. \end{aligned} \tag{6.42}$$

From the last equation, we find $w = 2$. Substituting into the second equation, we obtain $v = 1$. Then, the first equation gives $u = 1$.

This process involves *forward elimination* and *back substitution*. Forward elimination produces the pivots 2, −8, and 1 as in Eq. (6.42). It subtracts multipliers of each row from the rows beneath and reaches the "triangular" system. Back substitution, from bottom to top, substitutes each newly computed value into the equation above, and solve the system. For a system of n equations, the total number of operations for the forward elimination is

$$\sum_{1}^{n}(k^2 - k) = (1^2 + 2^2 + \cdots + n^2) - (1 + 2 + \cdots + n) = \frac{n^3 - n}{3}, \tag{6.43}$$

and for the back substitution, the number of operations is

$$1 + 2 + \cdots + n = \frac{n(n+1)}{2}. \tag{6.44}$$

Equation (6.40) can be written in the matrix form as

$$\mathbf{KU} = \mathbf{R}. \tag{6.45}$$

The whole process can then split into two steps. The first step is the *triangular factorization*. It gives

$$\mathbf{K} = \mathbf{LS}, \tag{6.46}$$

$$\mathbf{R} = \mathbf{LC}, \tag{6.47}$$

where $\mathbf{L}$ is lower triangular matrix, with 1s on the diagonal and the multipliers l_{ij} (taken from elimination) below the diagonal. $\mathbf{S}$ is the upper triangular matrix, which appears after forward elimination and before back substitution. The second step is then to solve

$$\mathbf{SU} = \mathbf{C}. \tag{6.48}$$

Since $\mathbf{S}$ is upper triangular matrix and the diagonal elements are the pivots in the Gauss elimination, $\mathbf{S}$ can be further written as $\mathbf{S} = \mathbf{D}\tilde{\mathbf{S}}$, where $\mathbf{D}$ is a diagonal matrix storing the diagonal elements of $\mathbf{S}$, i.e., $d_{ii} = S_{ii}$. For the classical continuum mechanics, when the $\mathbf{K}$ matrix is symmetric and the decomposition is unique, we obtain $\tilde{\mathbf{S}} = \mathbf{L}^t$. We thus have

$$\mathbf{DL}^t\mathbf{U} = \mathbf{C}, \tag{6.49}$$

and the $\mathbf{U}$ is obtained by a back substitution,

$$\mathbf{L}^t\mathbf{U} = \mathbf{D}^{-1}\mathbf{C}. \tag{6.50}$$

Since $\mathbf{D}$ is a diagonal matrix, there is no need of performing matrix inversion to obtain $\mathbf{D}^{-1}$. The elements of $\mathbf{D}^{-1}$ are the inverse of elements of $\mathbf{D}$.

When considering very large system, a direct method of solution can require a large amount of storage and computer time. The fact that considerable storage can be saved in an iterative solution has prompted a large amount of research effort to develop effective iterative schemes, such as Gauss–Seidel iteration method (Varga, 1962) and conjugate gradient method (Hestenes and Stiefel, 1952).

Procedures of Meshless Analysis of Elastic Static Problems

The step-by-step procedures of meshless analysis of elastostatics are given as follows.

Step 1. Read input data. Readers are referred to the user's manual (Appendix D).
Step 2. Generate shape functions $\mathbf{\Phi}$ and ψ and their derivatives $\mathbf{B}$ at all sampling points and at points on the boundary to specify the natural and essential boundary conditions (cf. Chapter 5).
Step 3. Form matrices $\mathbf{K}$ and $\mathbf{G}$ and forcing terms $\mathbf{F}$ and $\mathbf{f}$ (cf. Eqs. (5.45)–(5.48), (6.35)).
Step 4. Solve the governing equation (6.35) by Gauss elimination method to obtain the displacements $\mathbf{U}$ of all nodes and the Lagrange multipliers $\mathbf{\Lambda}$.
Step 5. The displacement field $\mathbf{u}$ and the strain field $\mathbf{e}$ for all sampling points are obtained according to Eqs. (5.32) and (5.33), respectively. The stresses are calculated according to Eqs. (6.5) and (6.7) and (6.13)–(6.15) for plane stress case and plane strain case, respectively.

Numerical Examples of Meshless Analysis of Elastic Static Problems

Plate with a Hole

The analytical solution of an infinite plate with a hole of radius a subjected to uniform tensile stress at infinity is given by

$$t_{xx} = T_x \left\{ 1 - \frac{a^2}{r^2} \left[\frac{3}{2} \cos(2\theta) + \cos(4\theta) \right] + \frac{3a^4}{2r^4} \cos(4\theta) \right\}, \tag{6.51}$$

$$t_{yy} = -T_x \left\{ \frac{a^2}{r^2} \left[\frac{1}{2} \cos(2\theta) - \cos(4\theta) \right] + \frac{3a^4}{2r^4} \cos(4\theta) \right\}, \tag{6.52}$$

$$t_{xy} = -T_x \left\{ \frac{a^2}{r^2} \left[\frac{1}{2} \sin(2\theta) - \sin(4\theta) \right] - \frac{3a^4}{2r^4} \sin(4\theta) \right\}, \tag{6.53}$$

$$u_r = \frac{T_x}{4\mu}\left\{ r\left[\frac{\kappa - 1}{2} + \cos(2\theta)\right] + \frac{a^2}{r}[1 + (1+\kappa)\cos(2\theta)] - \frac{a^4}{r^3}\cos(2\theta)\right\}, \tag{6.54}$$

$$u_\theta = \frac{T_x}{4\mu}\left[(1-\kappa)\frac{a^2}{r} - r - \frac{a^4}{r^3}\right]\sin\,(2\theta), \tag{6.55}$$

where μ is the shear modulus defined as

$$\mu \equiv \frac{E}{2(1+\upsilon)}, \tag{6.56}$$

and for plane strain,

$$\kappa = 3 - 4\upsilon, \tag{6.57}$$

for plane stress,

$$\kappa = \frac{(3-\upsilon)}{(1+\upsilon)}. \tag{6.58}$$

In this work, we specify the natural boundary conditions (stress free) around the hole ($r = a$) and the essential boundary conditions at $r = 5a$. 2400 nodes, as shown in Fig. 6.4, are employed. The analytical and numerical results for T_{xx} around the hole are shown in Fig. 6.5. It has been observed that the meshless solution and the analytical solution have a good agreement.

Plate with a Line Crack

An infinite plate with a line crack (along x-axis with crack size $= 2a$) as shown in Fig. 6.6 is subjected to tensile stress $\bar{t}_{yy} = \sigma_{yy} = \sigma$ where infinity is considered.

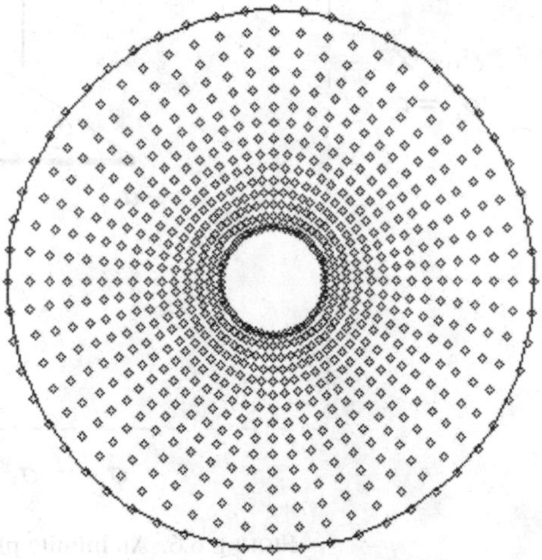

FIGURE 6.4. Meshless discretization of the plate with a hole.

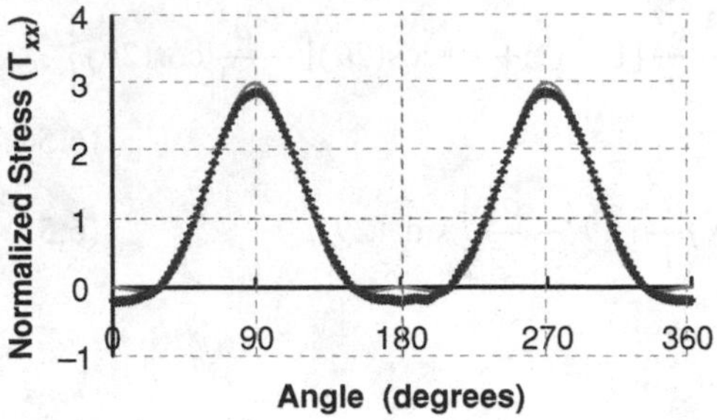

FIGURE 6.5. Numerical results of stress distribution. —, meshless results; —, analytical results.

The exact analytical solution is given by Sneddon and Lowengrub (1969) as

$$
\begin{aligned}
t_{xx} = {} & \frac{K_1}{\sqrt{arr_2}}\left[r_1\cos\left(\theta_1 - \frac{\theta+\theta_2}{2}\right) - \frac{r_1 a^2}{rr_2}\sin\theta_1\sin\frac{3}{2}(\theta+\theta_2)\right] \\
& + \frac{K_2}{\sqrt{arr_2}}\left[2r_1\sin\left(\theta_1 - \frac{\theta+\theta_2}{2}\right) - \frac{r_1 a^2}{rr_2}\sin\theta_1\cos\frac{3}{2}(\theta+\theta_2)\right] \\
& + \chi - \sigma,
\end{aligned} \tag{6.59}
$$

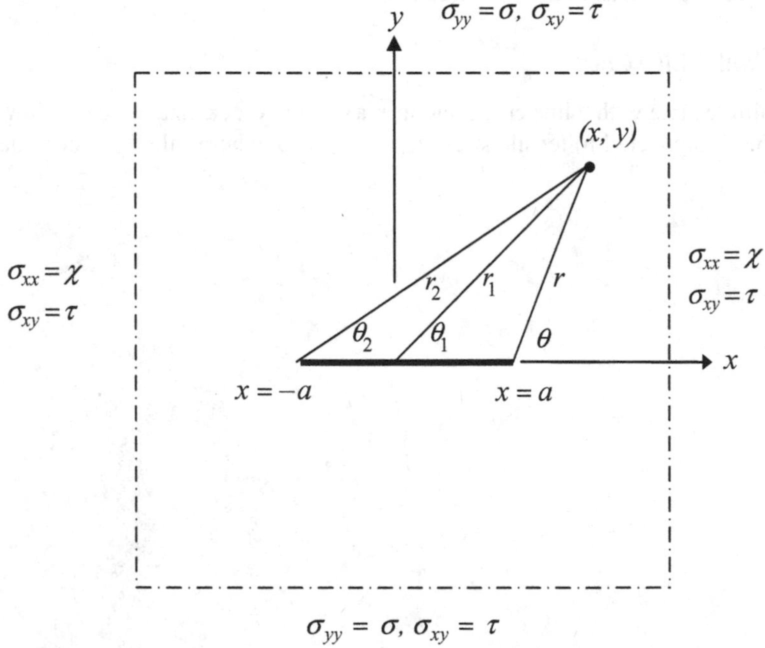

FIGURE 6.6. An infinite plate with a line crack.

$$t_{yy} = \frac{K_1}{\sqrt{arr_2}}\left[r_1 \cos\left(\theta_1 - \frac{\theta+\theta_2}{2}\right) + \frac{r_1 a^2}{rr_2}\sin\theta_1 \sin\frac{3}{2}(\theta+\theta_2)\right]$$
$$+ \frac{K_2}{\sqrt{arr_2}}\frac{r_1 a^2}{rr_2}\sin\theta_1 \cos\frac{3}{2}(\theta+\theta_2), \quad (6.60)$$

$$t_{xy} = \frac{K_1}{\sqrt{arr_2}}\frac{r_1 a^2}{rr_2}\sin\theta_1 \cos\frac{3}{2}(\theta+\theta_2)$$
$$+ \frac{K_2}{\sqrt{arr_2}}\left[r_1 \cos\left(\theta_1 - \frac{\theta+\theta_2}{2}\right) - \frac{r_1 a^2}{rr_2}\sin\theta_1 \sin\frac{3}{2}(\theta+\theta_2)\right], \quad (6.61)$$

$$4\mu u_x = \frac{K_1}{\sqrt{arr_2}}\left[(\kappa-1)rr_2 \cos\frac{\theta+\theta_2}{2} - 2r_1^2 \sin\theta_1 \sin\left(\theta_1 - \frac{\theta+\theta_2}{2}\right)\right]$$
$$+ \frac{K_2}{\sqrt{arr_2}}\left[(\kappa+1)rr_2 \sin\left(\frac{\theta+\theta_2}{2}\right) + 2r_1^2 \sin\theta_1 \cos\left(\theta_1 - \frac{\theta+\theta_2}{2}\right)\right]$$
$$- 0.5(\sigma-\chi)(\kappa+1)r_1\cos\theta_1, \quad (6.62)$$

$$4\mu u_y = \frac{K_1}{\sqrt{arr_2}}\left[(\kappa+1)rr_2 \sin\frac{\theta+\theta_2}{2} - 2r_1^2 \sin\theta_1 \cos\left(\theta_1 - \frac{\theta+\theta_2}{2}\right)\right]$$
$$+ \frac{K_2}{\sqrt{arr_2}}\left[(1-\kappa)rr_2 \cos\left(\frac{\theta+\theta_2}{2}\right) - 2r_1^2 \sin\theta_1 \sin\left(\theta_1 - \frac{\theta+\theta_2}{2}\right)\right]$$
$$- 0.5(\sigma-\chi)(\kappa-3)r_1\sin\theta_1, \quad (6.63)$$

where $K_1 = \sigma\sqrt{a}$ and $K_2 = \tau\sqrt{a}$.

The natural boundary conditions (stress free) along the crack surfaces and the essential boundary conditions along the boundary of the square of $0.2a \times 0.2a$ are specified according to the exact analytical solution (Sneddon and Lowengrub, 1969). One-hundred and twenty-six nodes, as shown in Fig. 6.7, are initially employed. With h and p refinements, we have used 232 nodes finally. The normalized stress t_{yy}/σ, which is the most critical crack opening stress, is plotted along the line crack in Fig. 6.8. The analytical results (Sneddon and Lowengrub, 1969) are also plotted in Fig. 6.8. The agreement between the analytical solution and the meshless solution is excellent. Also, it is noticed that the line crack is treated as a barrier that cuts off the nodal support according to the rule of visibility test (cf. Chapter 5).

In Fig. 6.9, the normalized stress t_{yy}/σ is displayed on the deformed shape of the specimen. The distribution of the meshless stress is in general quite close to that obtained by the finite element (FE) method. A region of compressive stress is seen at the right-hand side of the specimen due to the presence of a long crack, i.e., $a/w = 0.5$.

Three-Point Bending Beam with a Small Edge Crack

A concrete beam, 12 m in length, 3 m in height, 1 m in thickness, is simply supported at two ends and subjected to a concentrated load of 100 N at the center

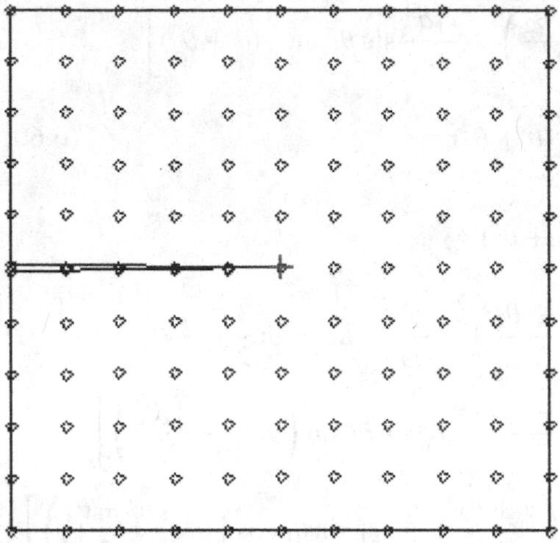

FIGURE 6.7. Meshless discretization of the cracked specimen.

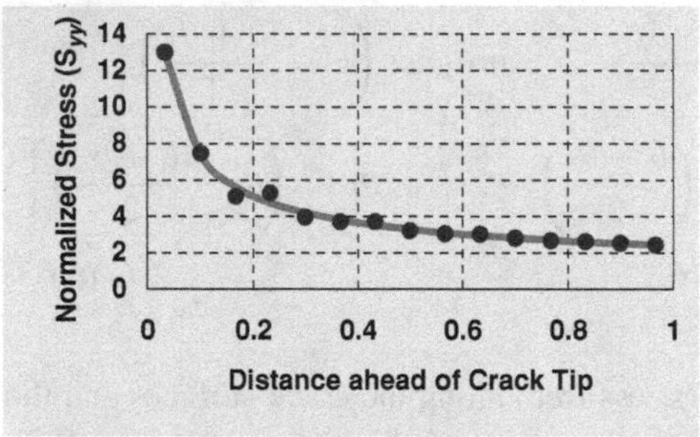

FIGURE 6.8. Numerical solution of the stress distribution along the line crack. • • •: meshless solution; —: exact analytical solution.

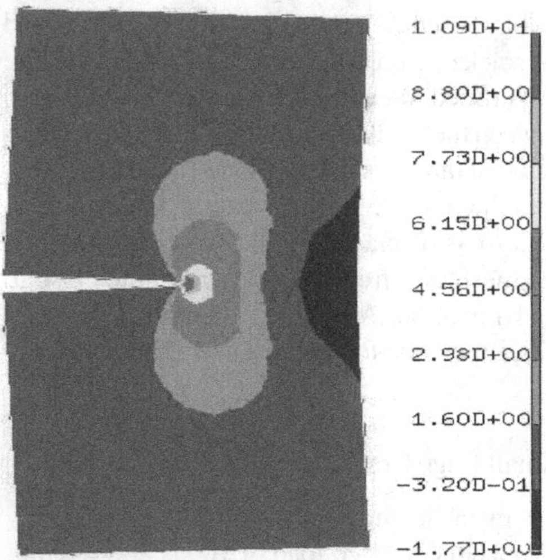

FIGURE 6.9. Stress distribution displayed on deformed shape.

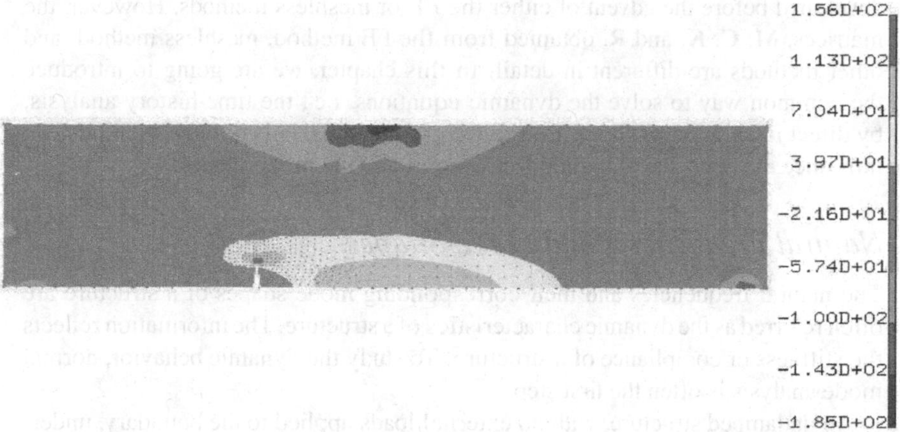

FIGURE 6.10. Stress (t_{xx} in GPa) distribution of a three-point bending beam with edge crack.

of the top surface. An edge crack extends from $\{x, y\} = \{4, 0.0\}$ to $\{4, 0.4\}$. The distribution of local meshless solution of t_{xx} is displayed in Fig. 6.10. It is seen that there is an intense and localized stress concentration around the crack tip, which is even larger than the maximum bending stress at the center of the bottom surface of the beam.

General Dynamic Problems

Wave Propagation Problems and Structural Dynamic Problems

Generally, dynamic problems can be categorized as either wave propagation problems or structural dynamic problems. In wave propagation problems, the loading is often an impact or an explosive blast. The structural responses are rich in high frequencies. In such problems, we are usually interested in the effects of stress wave and the transients produced. Thus, the time duration of analysis is usually short and is typically of the order of a wave traversal time across a structure. A problem that is not wave propagation problem, but for which inertia is important, is a structural dynamic problem. For structural dynamic problem, the frequency of excitation is usually of the same order as the structure's lowest natural frequencies of vibration.

Methods of structural dynamics are largely independent of the discretization methods. The standard form of dynamic problem can be written in a general form

$$\mathbf{M}\ddot{\mathbf{U}} + \mathbf{C}\dot{\mathbf{U}} + \mathbf{K}\mathbf{U} = \mathbf{R}, \tag{6.64}$$

where **M** is the mass matrix, **K** the stiffness matrix, **C** the damping matrix, and **R** the general force matrix. Many popular methods to solve it were formulated and developed before the advent of either the FE or meshless methods. However, the matrices, **M**, **C**, **K**, and **R**, obtained from the FE method, meshless method, and other methods are different in detail. In this chapter, we are going to introduce the common way to solve the dynamic equations, i.e., the time-history analysis, by direct integration methods. Before we go into the details of those methods, we introduce the concepts of natural frequencies and the mode shapes.

Natural Frequencies and Mode Shapes

The natural frequencies and their corresponding mode shapes of a structure are often referred as the dynamic characteristics of a structure. The information reflects the stiffness or compliance of a structure. To study the dynamic behavior, normal mode analysis is often the first step.

An undamped structure, with no external loads applied to the boundary, undergoes harmonic motion caused by initial conditions. Thus,

$$\mathbf{U} = \bar{\mathbf{U}} \sin \omega t, \tag{6.65}$$

and consequently

$$\ddot{\mathbf{U}} = -\omega^2 \bar{\mathbf{U}} \sin \omega t, \tag{6.66}$$

where $\bar{\mathbf{U}}$ is the amplitude of the vibrations and ω the circular frequency (radians per second).

Substituting $\mathbf{U}$ and $\ddot{\mathbf{U}}$ into Eq. (6.64) with $\mathbf{C}$ and $\mathbf{R}$ both being zero, we obtain

$$(\mathbf{K} - \omega^2 \mathbf{M})\bar{\mathbf{U}} = 0. \tag{6.67}$$

This is the basic statement of the vibration problem. Equation (6.67) is called a generalized eigenvalue problem. It has nontrivial solutions only if

$$\det(\mathbf{K} - \omega^2 \mathbf{M}) = 0. \tag{6.68}$$

Associated with each eigenvalue $\lambda_i = \omega_i^2$, there is an eigenvector $\bar{\mathbf{U}}_i$, which is called a normal mode. The lowest nonzero ω_i is called the fundamental vibration frequency. If $\mathbf{K}$ and $\mathbf{M}$ are $n \times n$ matrices, under conditions usually satisfied in structural analysis, Eq. (6.68) has n eigenvalues and n eigenvectors. All eigenvalues are positive if $\mathbf{K}$ and $\mathbf{M}$ are both positive definite. A partly or completely unsupported structure has positive semidefinite $\mathbf{K}$ and has one zero eigenvalue associated with each possible rigid-body motion.

Direct Integration Methods in Transient Analysis

In a time-history or dynamic response problem, we usually solve the dynamic equation in the form of Eq. (6.64) for $\mathbf{U}$, $\dot{\mathbf{U}}$, and $\ddot{\mathbf{U}}$ as functions of time. When $\mathbf{M}$, $\mathbf{C}$, and $\mathbf{K}$ are independent of $\mathbf{U}$, $\dot{\mathbf{U}}$, and $\ddot{\mathbf{U}}$, the system is linear. When $\mathbf{M}$, $\mathbf{C}$, and $\mathbf{K}$ are independent of time, the system is time invariant. If the material behavior is nonlinear, we use an internal force vector $\mathbf{R}^{\text{int}}$ replacing $\mathbf{KU}$ in Eq. (6.64), which is then called a nonlinear initial value problem.

There are two kinds of methods for time-history analysis: mode superposition method and direct integration method. For many structural dynamics or wave propagation problems, including those with complicated nonlinearities, direct integration is more expedient.

Methods of direct integration can be categorized as explicit or implicit. Explicit methods have the form

$$\mathbf{U}^{(n+1)} = \mathbf{f}(\mathbf{U}^{(n)}, \dot{\mathbf{U}}^{(n)}, \ddot{\mathbf{U}}^{(n)}, \mathbf{U}^{(n-1)}, \dot{\mathbf{U}}^{(n-1)}, \ddot{\mathbf{U}}^{(n-1)}, \ldots), \tag{6.69}$$

and hence permit $\mathbf{U}^{(n+1)}$ to be completely determined in terms of historical information consisting of displacements and time derivatives of displacements at time $n\Delta t$ and before. Implicit methods have the form

$$\mathbf{U}^{(n+1)} = \mathbf{f}(\dot{\mathbf{U}}^{(n+1)}, \ddot{\mathbf{U}}^{(n+1)}, \mathbf{U}^{(n)}, \dot{\mathbf{U}}^{(n)}, \ddot{\mathbf{U}}^{(n)}, \ldots), \tag{6.70}$$

and hence computation of $\mathbf{U}^{(n+1)}$ requires knowledge of the time derivatives of $\mathbf{U}^{(n+1)}$, which are unknown. Explicit and implicit methods have markedly different properties. The choices of method is strongly problem dependent. In this chapter, we introduce both explicit and implicit direct integration methods and consider only linear and time-invariant problems (cf. Bathe, 1996).

Central Difference Method

If the dynamic equation (6.64) is regarded as a system of ordinary differential equations with constant coefficients, it follows that any convenient finite difference expressions to approximate the accelerations and velocities in terms of displacements can be used. Therefore, a large number of differential finite difference expressions could be theoretically employed. One procedure that has been proved very effective in the solution of certain problems is the central difference method, in which it is assumed that

$$\ddot{\mathbf{U}}^{(n)} = (\mathbf{U}^{(n-1)} - 2\mathbf{U}^{(n)} + \mathbf{U}^{(n+1)})/(\Delta t)^2. \tag{6.71}$$

The error in the expansion, Eq. (6.71), is of order $(\Delta t)^2$, and to have the same order of error in the velocity expansion, we use

$$\dot{\mathbf{U}}^{(n)} = \frac{1}{2}(\mathbf{U}^{(n+1)} - \mathbf{U}^{(n-1)})/\Delta t. \tag{6.72}$$

The displacement solution for time $t + \Delta t$ is obtained by considering Eq. (6.64) at time t, i.e.,

$$\mathbf{M}\ddot{\mathbf{U}}^{(n)} + \mathbf{C}\dot{\mathbf{U}}^{(n)} + \mathbf{K}\mathbf{U}^{(n)} = \mathbf{R}^{(n)}. \tag{6.73}$$

Substituting the expressions of $\ddot{\mathbf{U}}^{(n)}$ and $\dot{\mathbf{U}}^{(n)}$ in Eqs. (6.71) and (6.72), respectively, into Eq. (6.73), we obtain

$$\begin{aligned}[\mathbf{M}/(\Delta t)^2 + \mathbf{C}/2\Delta t]\mathbf{U}^{(n+1)} = \mathbf{R}^{(n)} - [\mathbf{K} - 2\mathbf{M}/(\Delta t)^2]\mathbf{U}^{(n)} - [\mathbf{M}/(\Delta t)^2 \\ + \mathbf{C}/2\Delta t]\mathbf{U}^{(n-1)},\end{aligned} \tag{6.74}$$

from which we can solve $\mathbf{U}^{(n+1)}$. It should be noted that the solution of $\mathbf{U}^{(n+1)}$ is thus based on using the equilibrium conditions at time $t = n\Delta t$, i.e., Eq. (6.74). For this reason, the integration procedure is called an explicit integration method, and it is noted that such integration scheme do not require a factorization of the effective stiffness matrix in the step-by-step solution. On the other hand, the Houbolt, Wilson, and Newmark methods, discussed later, using the equilibrium conditions at time $t = (n+1)\Delta t$, are called implicit integration methods.

If central difference method is extremely effective when the damping matrix is zero and the mass matrix is diagonal, then the system of equations in Eq. (6.64) can be solved without factorizing a matrix. This can be achieved in both FE methods and meshless methods. It should be emphasized that the integration method requires that the time step Δt be smaller than a critical value, $\Delta t_{cr} = 2/\omega_{\max}$, where $\omega_{\max}$ is the largest frequency of the n degree of freedom system (cf. Eq. (6.68)). This requirement makes the central difference method a conditionally stable method.

Houbolt Method

The Houbolt integration scheme is somewhat related to the central difference method in which standard finite difference expressions are used to approximate the acceleration and velocity components in terms of the displacement components. The following finite difference expansions are employed in the Houbolt integration method (Houbolt, 1950):

$$\ddot{\mathbf{U}}^{(n+1)} = [2\mathbf{U}^{(n+1)} - 5\mathbf{U}^{(n)} + 4\mathbf{U}^{(n-1)} - \mathbf{U}^{(n-2)}]/(\Delta t)^2, \tag{6.75}$$

and

$$\dot{\mathbf{U}}^{(n+1)} = [11\mathbf{U}^{(n+1)} - 18\mathbf{U}^{(n)} + 9\mathbf{U}^{(n-1)} - 2\mathbf{U}^{(n-2)}]/6\Delta t, \tag{6.76}$$

which are two backward difference formulas with errors of order $(\Delta t)^2$.

In order to obtain the solution at time $t + \Delta t$, we now consider Eq. (6.64) at time $t + \Delta t$ (different from the central difference method), i.e.,

$$\mathbf{M}\ddot{\mathbf{U}}^{(n+1)} + \mathbf{C}\dot{\mathbf{U}}^{(n+1)} + \mathbf{K}\mathbf{U}^{(n+1)} = \mathbf{R}^{(n+1)}. \tag{6.77}$$

Substituting Eqs. (6.75) and (6.76) into Eq. (6.77) results

$$\begin{aligned}
[2\mathbf{M}/(\Delta t)^2 + 11\mathbf{C}/6\Delta t + \mathbf{K}]\mathbf{U}^{(n+1)} &= \mathbf{R}^{(n+1)} + [5\mathbf{M}/(\Delta t)^2 + 3\mathbf{C}/\Delta t]\mathbf{U}^{(n)} \\
&\quad - [4\mathbf{M}/(\Delta t)^2 + 3\mathbf{C}/2\Delta t]\mathbf{U}^{(n-1)} \\
&\quad + [\mathbf{M}/(\Delta t)^2 + \mathbf{C}/3\Delta t]\mathbf{U}^{(n-2)}.
\end{aligned} \tag{6.78}$$

As shown in Eq. (6.78), the solution of $\mathbf{U}^{(n+1)}$ requires knowledge of $\mathbf{U}^{(n)}$, $\mathbf{U}^{(n-1)}$, and $\mathbf{U}^{(n-2)}$. Although the knowledge of $\mathbf{U}^{(0)}$, $\dot{\mathbf{U}}^{(0)}$, and $\ddot{\mathbf{U}}^{(0)}$ is useful to start the Houbolt integration scheme, it is more accurate to calculate $\mathbf{U}^{(1)}$ and $\mathbf{U}^{(2)}$ by some other means; i.e., we employ special starting procedures. One way to proceed is to integrate Eq. (6.64) for the solutions of $\mathbf{U}^{(1)}$ and $\mathbf{U}^{(2)}$ by using a different integration scheme, possibly a conditionally stable method such as the central difference scheme with a fraction of Δt as the time step.

A basic difference between the Houbolt method and the central difference scheme is in the appearance of the stiffness matrix $\mathbf{K}$ in the matrix $2\mathbf{M}/(\Delta t)^2 + 11\mathbf{C}/6\Delta t + \mathbf{K}$ to be factorized to obtain the required displacements $\mathbf{U}^{(n+1)}$. The term $\mathbf{KU}^{(n+1)}$ appears because in (6.78) equilibrium is considered at time $t + \Delta t$ and not at time t as in the central difference method. The Houbolt method is, for this reason, an implicit integration scheme, whereas the central difference method was an explicit procedure. With regard to the time step Δt that can be used in the integration, there is no critical time step limit and Δt can in general be selected much larger than that given by the central difference method.

A noteworthy point is that the step-by-step solution scheme based on the Houbolt method reduces directly to a static analysis if mass and damping effects are neglected, whereas the central difference method could not be used. In other words, if $\mathbf{C} = \mathbf{M} = 0$, the Houbolt method yields the static solution for time-dependent loads.

Wilson θ Method

Let τ denote the increase in time, where $0 \leq \tau \leq \theta\Delta t$, then for the time interval $[t, t + \theta\Delta t]$, it is assumed that

$$\ddot{\mathbf{U}}(t + \tau) = \ddot{\mathbf{U}}(t) + \tau[\ddot{\mathbf{U}}(t + \theta\Delta t) - \ddot{\mathbf{U}}(t)]/\theta\Delta t. \tag{6.79}$$

Integrating Eq. (6.79), we obtain

$$\dot{\mathbf{U}}(t + \tau) = \dot{\mathbf{U}}(t) + \tau\ddot{\mathbf{U}}(t) + \tau^2[\ddot{\mathbf{U}}(t + \theta\Delta t) - \ddot{\mathbf{U}}(t)]/2\theta\Delta t, \tag{6.80}$$

and

$$\mathbf{U}(t + \tau) = \mathbf{U}(t) + \tau\dot{\mathbf{U}}(t) + \tau^2\ddot{\mathbf{U}}(t)/2 + \tau^3[\ddot{\mathbf{U}}(t + \theta\Delta t) - \ddot{\mathbf{U}}(t)]/6\theta\Delta t. \tag{6.81}$$

Using Eqs. (6.80) and (6.81), we have at time $t + \theta\Delta t$

$$\dot{\mathbf{U}}(t + \theta\Delta t) = \dot{\mathbf{U}}(t) + \theta\Delta t[\ddot{\mathbf{U}}(t + \theta\Delta t) + \ddot{\mathbf{U}}(t)]/2, \tag{6.82}$$

$$\mathbf{U}(t + \theta\Delta t) = \mathbf{U}(t) + \theta\Delta t\dot{\mathbf{U}} + (\theta\Delta t)^2[\ddot{\mathbf{U}}(t + \theta\Delta t) + 2\ddot{\mathbf{U}}(t)]/6, \tag{6.83}$$

from which we can solve $\ddot{\mathbf{U}}(t + \theta\Delta t)$ and $\dot{\mathbf{U}}(t + \theta\Delta t)$ in terms of $\mathbf{U}(t + \theta\Delta t)$:

$$\ddot{\mathbf{U}}(t + \theta\Delta t) = 6[\mathbf{U}(t + \theta\Delta t) - \mathbf{U}(t)]/(\theta\Delta t)^2 - 6\dot{\mathbf{U}}(t)/\theta\Delta t - 2\ddot{\mathbf{U}}(t), \tag{6.84}$$

$$\dot{\mathbf{U}}(t + \theta\Delta t) = 3[\mathbf{U}(t + \theta\Delta t) - \mathbf{U}(t)]/\theta\Delta t - 2\dot{\mathbf{U}}(t) - \theta\Delta t\ddot{\mathbf{U}}(t)/2. \tag{6.85}$$

To obtain the solution for the displacements, velocities, and accelerations at time $t + \Delta t$, the equilibrium equation, Eq. (6.64), is considered at time $t + \theta\Delta t$. However, because the accelerations are assumed to vary linearly, a linearly extrapolated load vector is used; i.e., the equation employed is

$$\mathbf{M}\ddot{\mathbf{U}}(t + \theta\Delta t) + \mathbf{C}\dot{\mathbf{U}}(t + \theta\Delta t) + \mathbf{KU}(t + \theta\Delta t) = \mathbf{R}(t + \theta\Delta t), \tag{6.86}$$

where

$$\mathbf{R}(t + \theta\Delta t) = \mathbf{R}(t) + \theta[\mathbf{R}(t + \theta\Delta t) - \mathbf{R}(t)]. \tag{6.87}$$

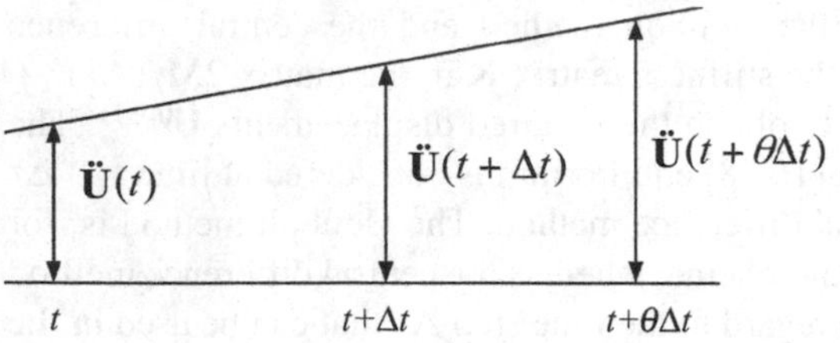

FIGURE 6.11. Linear acceleration assumption of Wilson θ method.

Substituting Eqs. (6.84) and (6.85) into Eq. (6.86), an equation is obtained from which $\mathbf{U}(t+\theta\Delta t)$ can be solved. Substitute $\mathbf{U}(t+\theta\Delta t)$ into Eqs. (6.84) and (6.85) to obtain $\ddot{\mathbf{U}}(t+\theta\Delta t)$ and $\dot{\mathbf{U}}(t+\theta\Delta t)$. Then, from Eqs. (6.79) to (6.81), we obtain $\mathbf{U}(t+\Delta t)$, $\dot{\mathbf{U}}(t+\Delta t)$, and $\ddot{\mathbf{U}}(t+\Delta t)$ based on $\tau = \Delta t$. Interested reader may find the complete algorithm of Wilson θ method (Fig. 6.11) from Bathe's book (1996).

As it can be seen that the Wilson θ method is also an implicit integration method, the stiffness matrix $\mathbf{K}$ is a coefficient matrix to the unknown displacement vector. It may also be noted that no special starting procedures are needed since the displacements, velocities, and accelerations at time $t+\Delta t$ are expressed in terms of the same quantities at time t only.

Newmark Method

The Newmark integration scheme can also be understood to be an extension of linear acceleration method. The following assumptions are used (Newmark, 1959):

$$\dot{\mathbf{U}}(t+\Delta t) = \dot{\mathbf{U}}(t) + [(1-\delta)\ddot{\mathbf{U}}(t) + \delta\ddot{\mathbf{U}}(t+\Delta t)]\Delta t, \tag{6.88}$$

$$\mathbf{U}(t+\Delta t) = \mathbf{U}(t) + \Delta t\dot{\mathbf{U}} + [(0.5-\alpha)\ddot{\mathbf{U}}(t) + \alpha\ddot{\mathbf{U}}(t+\Delta t)](\Delta t)^2, \tag{6.89}$$

where α and δ are parameters that can be determined to obtain integration accuracy and stability. When $\delta = \frac{1}{2}$ and $\alpha = \frac{1}{6}$, Eqs. (6.88) and (6.89) correspond to the linear acceleration method, which is also obtained by using $\theta = 1$ in the Wilson θ method. Newmark originally proposed as an unconditionally stable scheme: the constant-average-acceleration method, in which case $\delta = \frac{1}{2}$ and $\alpha = \frac{1}{4}$.

In addition to Eqs. (6.88) and (6.89), for the solution of displacements, velocities, and accelerations at time $t+\Delta t$, the equilibrium equation (6.64) at time $t+\Delta t$ is also considered:

$$\mathbf{M}\ddot{\mathbf{U}}(t+\Delta t) + \mathbf{C}\dot{\mathbf{U}}(t+\Delta t) + \mathbf{K}\mathbf{U}(t+\Delta t) = \mathbf{R}(t+\Delta t). \tag{6.90}$$

Solving from Eq. (6.90) for $\ddot{\mathbf{U}}(t+\Delta t)$ in terms of $\mathbf{U}(t+\Delta t)$ and then substituting $\ddot{\mathbf{U}}(t+\Delta t)$ into Eq. (6.88), we obtain equations for $\dot{\mathbf{U}}(t+\Delta t)$ and $\ddot{\mathbf{U}}(t+\Delta t)$, each in terms of the unknown displacements $\mathbf{U}(t+\Delta t)$ only. These two relations $\dot{\mathbf{U}}(t+\Delta t)$ and $\ddot{\mathbf{U}}(t+\Delta t)$ are substituted into Eq. (6.90), to solve $\mathbf{U}(t+\Delta t)$, afterward, using Eqs. (6.88) and (6.89), $\dot{\mathbf{U}}(t+\Delta t)$ and $\ddot{\mathbf{U}}(t+\Delta t)$ can also be calculated. Interested reader may find the complete algorithm of Newmark's method (Fig. 6.12) from Bathe's book (1996).

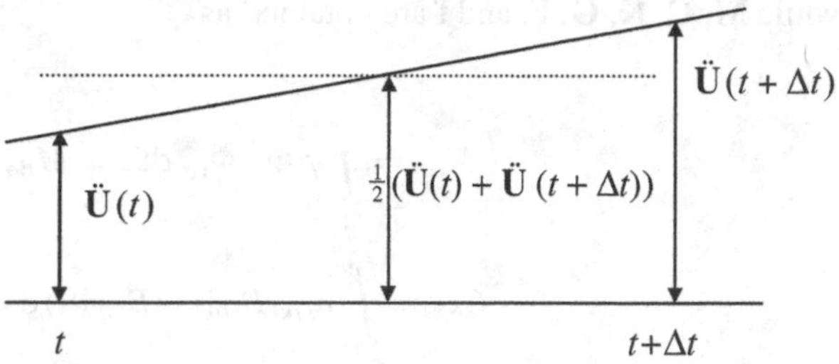

FIGURE 6.12. Newmark's constant-average acceleration scheme.

The close relationship between the computer implementation of the Newmark and the Wilson θ method should be noted.

With explicit methods of direct integration, stability typically requires that the time step be small enough that information does not propagate more than one element per time. Explicit methods are ideal for wave propagation problems in which behavior at the stress wave front is of engineering importance. Here the stability restriction is not a serious disadvantage because a small Δt is necessary for accuracy. Other factors in favor of explicit time integration are easy implementation, accurate treatment of general nonlinearity, and the capability of treating very large problems with only modest computer storage requirements. For structural dynamic problems, time scales and analysis durations are usually long and accuracy consideration alone would permit a Δt much larger than the upper limit of Δt for stable explicit integration. Explicit methods are then not as well suited to this class of problems, as they are suited for wave propagation problems.

The only advantage of implicit methods over explicit methods is that they allow a much large Δt because they are unconditionally stable. Implicit methods are expensive for wave propagation problems since accuracy requires a small Δt. For long-duration structural dynamic problems, implicit methods are usually more effective than explicit methods.

Meshless Analysis of Elastodynamic Problems

Meshless Formulation of Elastodynamics

The meshless formulation of dynamic problem has been obtained in Chapter 5. The governing equations for the displacements can be expressed as

$$\bar{\mathbf{K}}\mathbf{M}\ddot{\mathbf{U}} + \bar{\mathbf{K}}\mathbf{C}\dot{\mathbf{U}} + \mathbf{U} = \bar{\mathbf{K}}\mathbf{F} + \bar{\mathbf{G}}\mathbf{f}, \tag{6.91}$$

where $\bar{\mathbf{K}}^t$ and $\bar{\mathbf{G}}^t$ can be obtained by solving the following system of linear equations

$$\begin{vmatrix} \mathbf{K}^t & \mathbf{G} \\ \mathbf{G}^t & 0 \end{vmatrix} \begin{vmatrix} \bar{\mathbf{K}}^t \\ \bar{\mathbf{G}}^t \end{vmatrix} = \begin{vmatrix} \mathbf{I} \\ 0 \end{vmatrix}, \tag{6.92}$$

while **M**, **C**, **K**, **G**, **F**, and **f** are obtained as

$$M_{\alpha\beta} = \int_{\Omega} \rho \Phi_{i\beta} \Phi_{i\alpha} \, d\Omega = M_{\beta\alpha}, \tag{6.93}$$

$$C_{\alpha\beta} = \int_{\Omega} a_{ijkl} B_{ij\alpha}(\mathbf{x}) B_{kl\beta}(\mathbf{x}) \, dV, \tag{6.94}$$

$$K_{\alpha\beta} = \int_{\Omega} A_{ijkl} B_{ij\alpha}(\mathbf{x}) B_{kl\beta}(\mathbf{x}) dV, \tag{6.95}$$

$$G_{\alpha\beta} = \int_{\Gamma_u} \Phi_{i\alpha} \Psi_{i\beta} \, dS, \tag{6.96}$$

$$F_{\alpha} = \int_{\Omega} \rho f_j \Phi_{j\alpha} \, dV + \int_{\Gamma_t} \bar{t}_j \Phi_{j\alpha} \, dS, \tag{6.97}$$

$$f_{\alpha} = \int_{\Gamma_t} \Psi_{i\alpha} \bar{u}_i \, dS. \tag{6.98}$$

It is obvious that Eq. (6.91) can be rewritten in the same form as Eq. (6.64), i.e.,

$$\hat{\mathbf{M}}\ddot{\mathbf{U}} + \hat{\mathbf{C}}\dot{\mathbf{U}} + \hat{\mathbf{K}}\mathbf{U} = \hat{\mathbf{R}}, \tag{6.99}$$

where $\hat{\mathbf{M}} = \bar{\mathbf{K}}\mathbf{M}$, $\hat{\mathbf{C}} = \bar{\mathbf{K}}\mathbf{C}$, $\hat{\mathbf{K}} = \mathbf{I}$, and $\hat{\mathbf{R}} = \bar{\mathbf{K}}\mathbf{F} + \bar{\mathbf{G}}\mathbf{f}$. From now on we can employ the direct integration methods introduced in the section "General Dynamic Problems" to solve the meshless dynamic equations.

Procedures of Meshless Analysis of Elastic Dynamic Problems

The step-by-step procedures of meshless analysis of elastodynamics are given as follows.

Step 1. Read input data. Readers are referred to the user's manual (Appendix D).
Step 2. Generate shape functions $\mathbf{\Phi}$, ψ, and their derivatives **B** at all sampling points and at points on the boundary to specify the natural and essential boundary conditions (cf. Chapter 5).
Step 3. Form matrices **M, C, K**, and **G** and forcing terms **F** and **f** (cf. Eqs. (6.93)–(6.98)).
Step 4. Solve Eq. (6.92) for $\bar{\mathbf{K}}^t$ and $\bar{\mathbf{G}}^t$.

Step 5. Calculate the generalized mass, damping, stiffness matrices, and forcing term as

$$\begin{aligned} \hat{\mathbf{M}} &= \bar{\mathbf{K}}\mathbf{M}, \\ \hat{\mathbf{C}} &= \bar{\mathbf{K}}\mathbf{C}, \\ \hat{\mathbf{K}} &= \mathbf{I}, \\ \mathbf{F}^{\text{ext}} &= \hat{\mathbf{R}} = \bar{\mathbf{K}}\mathbf{F} + \bar{\mathbf{G}}\mathbf{f}. \end{aligned} \tag{6.100}$$

Step 6. Solve the governing equation (6.99) with the specified initial conditions $\mathbf{U}(t=0) = \mathbf{U}^0$ and $\dot{\mathbf{U}}(t=0) = \mathbf{V}^0$ by using the one of the direct integration methods (Newmark method is employed in the disk) to obtain the displacements $\mathbf{U}$ of all nodes.

Step 7. The displacement field $\mathbf{u}$ and the strain field $\mathbf{e}$ as functions of time t for all sampling points are obtained according to Eqs. (5.32) and (5.33), respectively. The stresses are calculated according to Eqs. (6.5) and (6.7) and (6.13)–(6.15) for plane stress case and plane strain case, respectively.

Numerical Examples of Meshless Analysis of Elastodynamic Problems

Shown in Fig. 6.13 is a cracked specimen with width w and height $2h$. The horizontal edge crack extends from $x = 0$ to $x = a$. A uniformly distributed normal stress, $t_{yy} = \sigma(t)$, is applied along the top and bottom surfaces ($y = \pm h$). A plane strain condition is assumed.

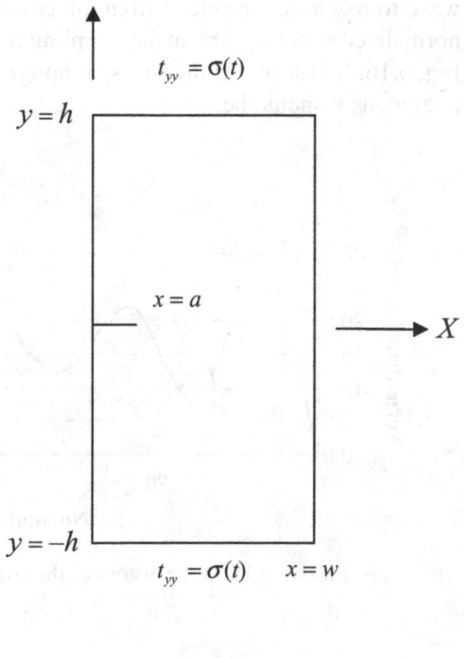

FIGURE 6.13. Edge cracked specimen subjected to mode-I tensile stress.

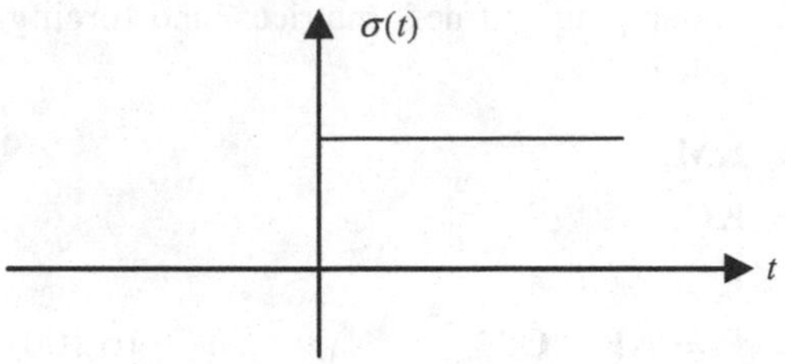

FIGURE 6.14. The applied loading as a function of time.

Case 1. The applied loading is a Heaviside step function in time as shown in Fig. 6.14, i.e., $\sigma(t) = \sigma^0$ for $t \geq 0$. A large damping coefficient is incorporated. Newmark method is employed for the direct integration. The normalized stress, t_{yy}/σ^0, at the sampling point nearest to the crack tip is plotted as a function of normalized time, t/τ, in Fig. 6.15 while τ is defined as $w/\sqrt{E/\rho}$. It is seen that as time becomes larger the stresses approach the static values. It indicates the validity of the computer software and, also, verifies that the time step, Δt, used in this work is small enough to yield accurate solutions.

Case 2. The applied loading in this case is expressed as

$$t_{yy}(x, \pm h, t) = \begin{cases} \sigma^0 \sin(2\pi t/t_0), & \text{if } t \leq t_0 \\ 0, & \text{if } t > t_0 \end{cases}$$

where $t_0 = 0.5\tau$ is small comparing with the time needed for the longitudinal wave to reach the line crack from the edges where the loadings are applied. The normalized stress, t_{yy}/σ^0, at the sampling point nearest to the crack tip is plotted in Fig. 6.16. It is seen that the stress, although still oscillating, is reducing gradually to zero, as it should be.

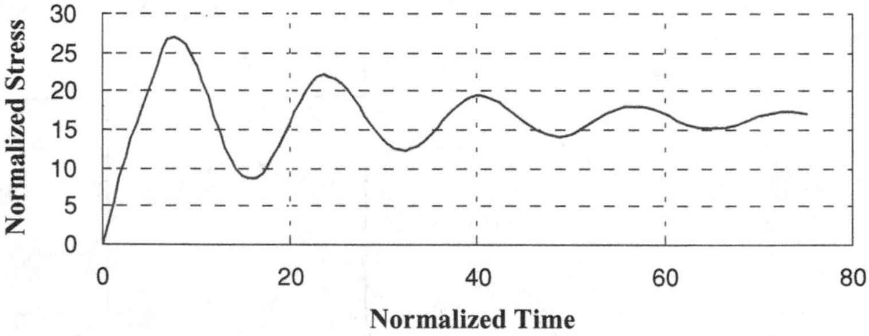

FIGURE 6.15. Time history of the stress near the crack tip (Case 1).

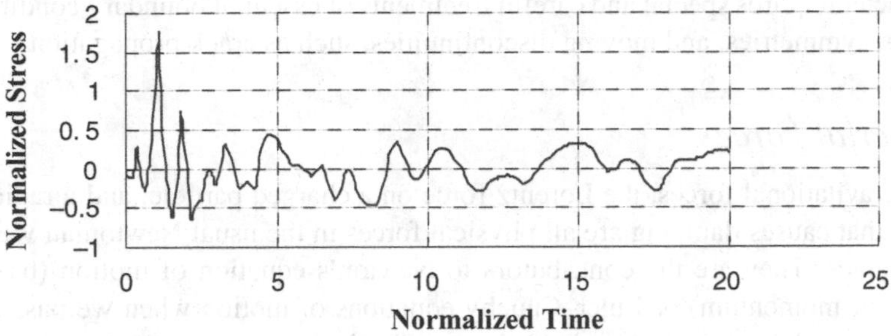

FIGURE 6.16. Time history of the stress near the crack tip (Case 2).

Meshless Analysis of Multiphase Materials

Multiphase Materials

Industry materials are generally inhomogeneous. They usually consist of microstructure and have multiphases, examples including ceramics, polycrystalline solids, concretes, etc. The boundaries between phases are usually irregular and random. Some of materials usually consist of defects such as microcracks, voids, and dislocations. Therefore, to say the least, it is very difficult to model and to perform numerical simulation of the detailed feature of multiphase materials. On the other hand, digital imaging data from CT, ultrasound, MRI, etc., are abounding.

CT, also known as CAT scanning or X-ray-computed topography, is a completely nondestructive technique that enables one to visualize detailed features in the interior of opaque solid objects and to obtain information on their 3D geometry and composition. In CT, cross-sectional images are generated by projecting a thin beam of X-ray through one plane of an object from many different angles. A 2D image of a section or a slice of a 3D object usually has 512×512 pixels. The value of each pixel is a measure of the reduction in X-ray intensity and energy, which in turn is a measure of the density of the material at that point. Therefore, the values at the pixels can be taken as the input to specify the configuration and composition of the specimen.

Meshless methods can be constructed solely in terms of nodes without the need of a highly structured mesh as required in FE method. In a meshless method, the approximation of any scalar-valued function, $\tilde{U}(\mathbf{x})$, can be expressed as an inner product between a vector of shape functions, $\mathbf{\Phi}(\mathbf{x})$, and a vector of nodal values, $\mathbf{U}$, as

$$\tilde{U}(\mathbf{x}) = \mathbf{\Phi}(\mathbf{x}) \cdot \mathbf{U}, \tag{6.101}$$

which has the same form as in the FE method. However, there is a characteristic difference between FE method and meshless method: Eq. (6.101) is an approximation rather than an interpolation, i.e., in meshless method, $\tilde{U}(\mathbf{x}_i) \neq U_i$. This character requires special and careful treatments of essential boundary conditions, mirror symmetries, and moving discontinuities, such as crack propagation.

Material Forces

The gravitational forces, the Lorentz force on a charged particle, and a radiation force that causes damping are all physical forces in the usual Newtonian view of mechanics. They are the contributors to Newton's equation of motion (balance of linear momentum) or Euler–Cauchy equations of motion when we pass from discrete model to continuum field theory. Physical forces are generated by displacements in physical space. For a continuous body, this means a change in its actual position in its physical configuration at time t (Maugin, 1995).

On the other hand, the concept of material forces was first introduced by Eshelby (1951), elaborated and further developed by Maugin (1993, 1995). Material forces are generated by displacement, not in physical space, but on material manifold. For example, they can be generated by (a) an infinitesimal rigid displacement of a finite region surrounding a point of singularity in an elastic body (Eshelby, 1951), (b) an infinitesimal displacement of a dislocation line (Peach and Koehler, 1950), (c) an infinitesimal increase in the length of a crack (Casal, 1978; Rice, 1968). This characteristic property of material forces also leads to their christening as inhomogeneity forces. Material inhomogeneity is defined as the dependence of properties (not the solution), such as density, elastic coefficients, viscosity, and plasticity threshold, on the material point. These inhomogeneities may be more or less continuous such as in metallurgically superficially treated specimens or in a polycrystal observed at a mesoscopic scale, or it may change abruptly such as in laminated composite or in a body with foreign inclusions or cavities.

For thermoelastic material, the governing equations of material forces may be expressed as

$$B_{KL,K} + F_L = \dot{P}_L, \tag{6.102}$$

where the pseudomomentum **P**, Eshelby stress **B**, and material force **F** are derived to be (Maugin, 1995):

$$P_L \equiv -\rho^0 v_k x_{k,L}, \tag{6.103}$$

$$B_{KL} = -(K - W)\delta_{KL} - T_{KM}C_{LM}, \tag{6.104}$$

$$\begin{aligned} F_L = -\rho^0 f_l x_{l,L} + \frac{1}{2} v_k v_k \left(\rho^0\right)_L + \left(\rho^0 \gamma T / T^0 + a_{KM} E_{KM}\right) T_L \\ + (\rho^0 \gamma)_L \frac{T^2}{2T^0} + a_{KM,L} E_{KM} T - \frac{1}{2} A_{IJMN,L} E_{IJ} E_{MN}. \end{aligned} \tag{6.105}$$

It is seen that the material force in thermoelastic solid is due to (1) body force **f**, (2) temperature gradient ∇T, and (3) the material inhomogeneities in density ρ^0 and all the thermoelastic coefficients γ, **a**, and **A**. In Eqs. (6.103)–(6.105), **v, C, E**, K, W, and T^0 are the velocity, Green deformation tensor, Lagrangian Strain, kinetic energy, strain energy, and reference temperature, respectively. It should be emphasized that it is almost impossible and even erroneous to calculate the derivatives of the material properties through FE method, and, on the other hand, it is natural and easy to do so through meshless method.

Also, for 2D problems in the presence of propagating crack, the material force associated with the crack tip is obtained as

$$\mathbf{F} = \lim_{\Gamma \to 0} \int_{\Gamma} \mathbf{N} \cdot (\mathbf{V} \otimes \mathbf{P} - \mathbf{B})\, d\Gamma, \tag{6.106}$$

where Γ denotes the cross-sectional circuit around the crack tip; $\mathbf{N}$ the unit vector normal to Γ pointing away from the crack tip, and $\mathbf{V}$ the velocity of crack propagation. Notice that crack propagation is a movement on material manifold, not in physical space, therefore, $\mathbf{V}$ is not equal to the material time rate of change of the position vector (velocity) of any particle. It can be shown that, in a very special case, the projection of $\mathbf{F}$ in the direction tangent to the crack path behind the crack tip is reduced to the J-integral, which is path independent if the material within Γ is homogeneous.

For readers interested in material forces, more recent works are found in Chen and Lee (2005), Lee et al. (2004), Lee and Chen (2005).

Crack Propagation

In 2D fracture problems, mode I fracture may lead to self-similar crack extension due to symmetry. In general case, especially in case of multiphase material, we encounter mixed mode fracture problems. Therefore, to determine the direction of crack extension is an unavoidable task. Usually, we use the maximum opening stress criterion or the maximum energy release rate criterion to determine the direction of crack propagation. For example, using maximum opening stress criterion, the current crack tip will extends to $\{r_c, \theta\}$ if the opening stress $t_{\theta\theta}$ is maximum at $\{r_c, \theta\}$, where $r_c > 0$ is small and finite constant. One may consider that $t_{\theta\theta}(r_c, \theta)$ is the driving force distributed along an arc with a radius r_c with respect to the current crack tip. If the material is homogeneous, the maximum opening stress criterion is reasonable, i.e., the information of driving force is enough to determine the direction of crack extension. However, if the material is inhomogeneous, one has to consider the resistance, i.e., the toughness, distributed in front of crack tip. In this work, we propose that the current crack tip will extend to $\{r_c, \theta\}$ if the ratio

$$R(r_c, \theta) \equiv \frac{t_{\theta\theta}(r_c, \theta)}{t_c(r_c, \theta)}, \tag{6.107}$$

reaches a maximum at $\{r_c, \theta\}$, where t_c is the toughness associated with the opening stress. Crack propagation process can be viewed as a changing of crack tip with a moving barrier following the advancing of the crack tip. It is noticed that meshless analysis of crack propagation does not involve the formidable task of constantly remeshing the cracking specimen. It only needs the updating of the barrier and the sprinkle of additional nodes in front of the current crack tip to enhance the solution accuracy.

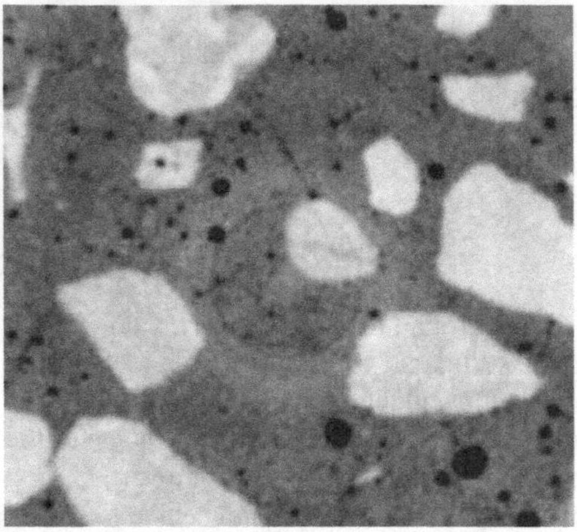

FIGURE 6.17. X-ray image of concrete indicting aggregate, paste, and voids.

Numerical Examples

A typical X-ray image of a slice of concrete and its meshless computational model are shown in Figs. 6.17 and 6.18, respectively. As can be seen from Fig. 6.17, there are aggregates, pastes, and voids, which are reproduced in the meshless model, Fig. 6.18. Both Figs. 6.17 and 6.18 are in terms of points (pixels). The only difference is the resolution. Here, for the sake of computational efficiency, we use a model with reduced resolution for the meshless computation. It turns out that, upon comparisons of numerical results, there is no significant difference by using models with reduced resolution.

The specimen is subjected to compression at the top and bottom edges, i.e., in the Y-direction, with a specified displacement of 1% of the height of the specimen. Stress distributions are displayed in Fig. 6.19(a)–(d). It can be seen from Fig. 6.19(a)–(d) that there is a nonuniform distribution of stresses. This is because

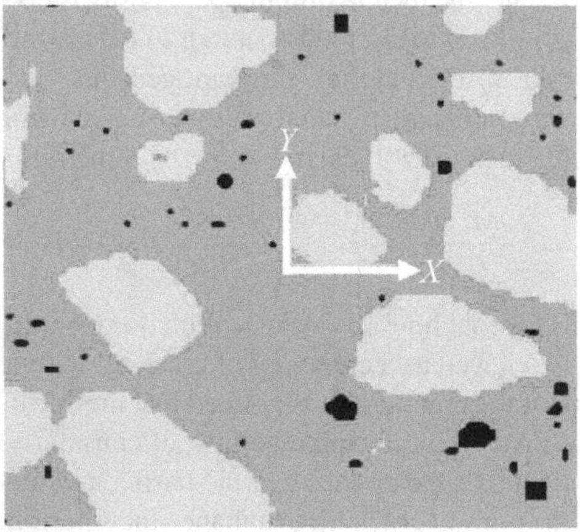

FIGURE 6.18. Meshless model of concrete with reduced resolution. Yellow: aggregate, blue: paste, black: voids.

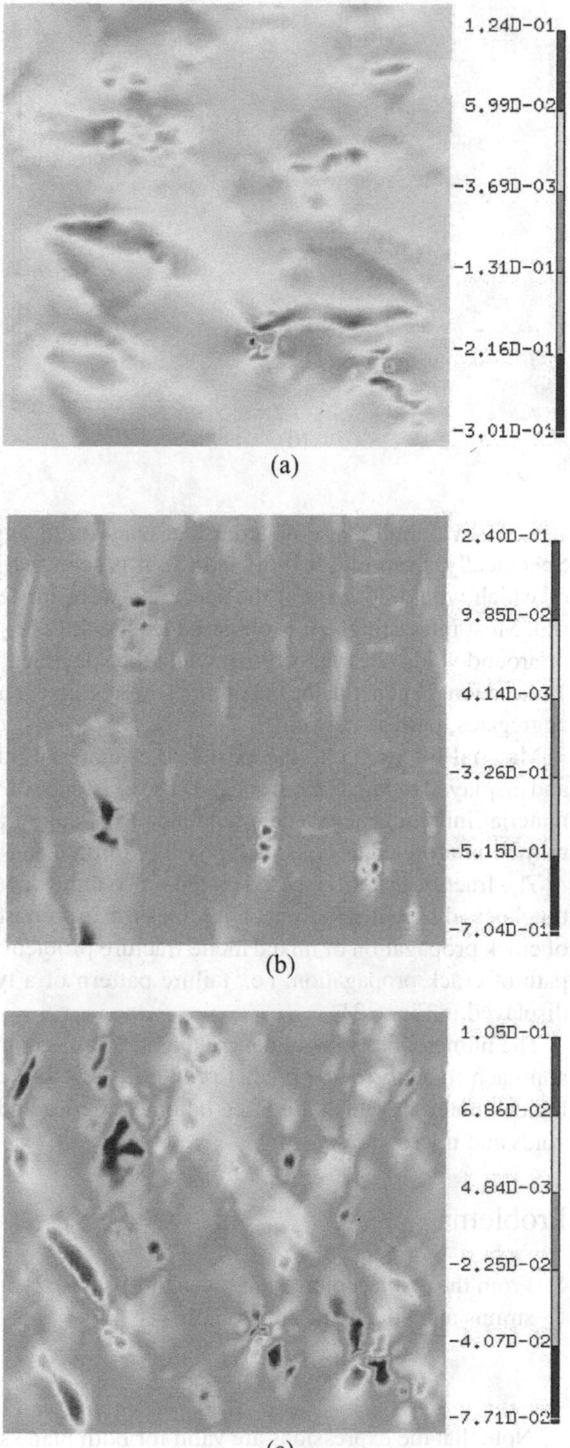

FIGURE 6.19. (a) Distribution of Cauchy stress σ_{xx}. (b) Distribution of Cauchy stress σ_{yy}. (c) Distribution of Cauchy stress σ_{xy}. (d) Distribution of von Mises stress.

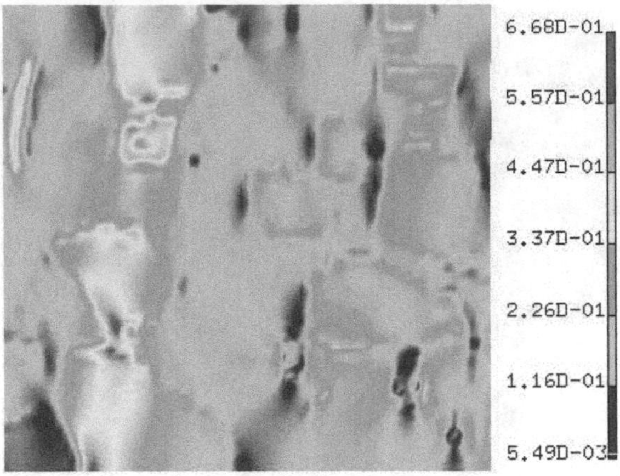

(d)

FIGURE 6.19. (*Continued*)

concrete is a multiphase material that consists of aggregates, pastes, and voids. Specifically, from Fig. 6.19(a) and (c), it is seen that there are σ_{xx} and σ_{xy}, both of which would not exist if the specimen were treated as a homogeneous material. Most interestingly, it is observed tensile stress σ_{yy} appears in the region near or around voids (cf. Fig. 6.19(b)) while the specimen is under compression in the Y-direction. In general, the maximum stresses appear around the interface between aggregates, pastes, and voids.

Material forces due to the existence of material inhomogeneity are calculated and displayed in Fig. 6.20(a) and (b). The existence of material forces indicates the material inhomogeneity and a tendency to change the material manifold, which may be interpreted as crack will possibly initiate and propagate.

The fracture criterion, based on the ratio of the opening stress over the material toughness distributed in front of the crack tip, is proposed to determine the direction of crack propagation of mixed mode fracture problem in multiphase material. The path of crack propagation, i.e., failure pattern of a typical concrete specimen is displayed in Fig. 6.21.

The numerical results of concrete material has demonstrated a very promising approach to analyze and predict properties and life of various industry and biological materials which are often inhomogeneous and have multiphase including voids and microcracks.

Problems

1. From the displacement field given in Eqs. (6.62) and (6.63), find the following strains analytically or numerically

$$e_{xx} = \frac{\partial u_x}{\partial x}, \quad e_{yy} = \frac{\partial u_y}{\partial y}, \quad \gamma_{xy} = 2e_{xy} = \frac{\partial u_x}{\partial y} + \frac{\partial u_y}{\partial x}.$$

Note that the expressions are valid for both plane strain and plane stress cases.

FIGURE 6.20. (a) Distribution of material force f_x. (b) Distribution of material force f_y.

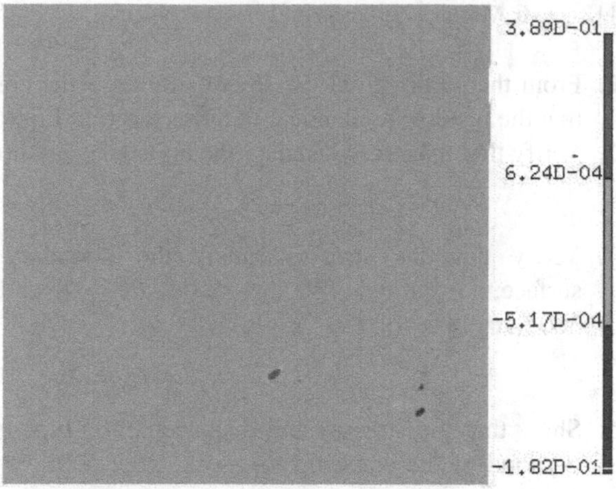

(a)

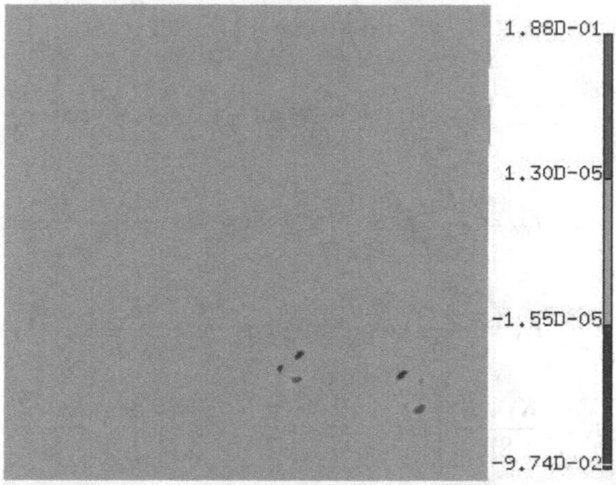

(b)

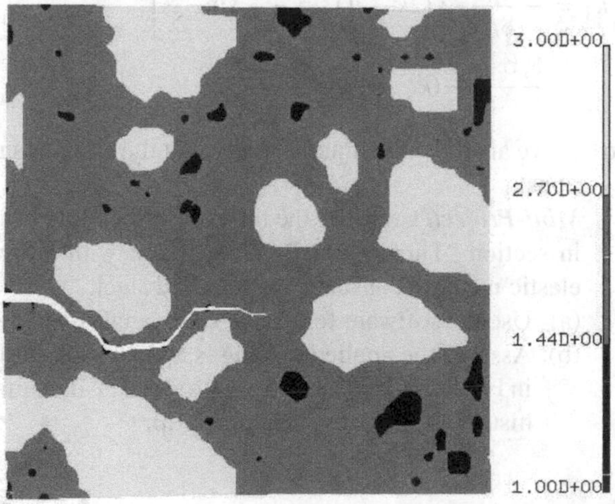

FIGURE 6.21. Failure (crack growth) pattern of concrete predicted by meshless method.

2. From the strains, find the stresses through the constitutive equations. Verify that the results are identical to those given in Eqs. (6.59)–(6.61).
3. Verify that the stresses satisfy the boundary conditions at infinity, i.e.,

$$r = r_1 = r_2 \to \infty, \quad t_{xx} = \tau, \quad t_{yy} = \sigma, \quad t_{xy} = \chi.$$

4. Verify that the stresses satisfy the boundary conditions at the crack surface, i.e., at $\theta = 180^\circ$, $\theta_2 = 0^\circ$, $\theta_1 = 0^\circ$ or 180° (or $\theta = -180^\circ$, $\theta_2 = 360^\circ$, $\theta_1 = 0^\circ$ or 180°),

$$t_{yy} = t_{xy} = 0.$$

5. Show that the stresses and displacements, Eqs. (6.59)–(6.63), in the neighborhood of the crack tip ($\theta_1 \to 0, \quad \theta_2 \to 0, \quad r_1 \to a, \quad r_2 \to 2a$) can be expressed as

$$\begin{aligned} t_{xx} = & \frac{K_1}{\sqrt{2r}} \cos\frac{\theta}{2}\left(1 - \sin\frac{\theta}{2}\sin\frac{3\theta}{2}\right) \\ & - \frac{K_2}{\sqrt{2r}} \sin\frac{\theta}{2}\left(2 + \cos\frac{\theta}{2}\cos\frac{3\theta}{2}\right) + \chi - \sigma + \cdots \end{aligned}$$

$$t_{yy} = \frac{K_1}{\sqrt{2r}} \cos\frac{\theta}{2}\left(1 + \sin\frac{\theta}{2}\sin\frac{3\theta}{2}\right) + \frac{K_2}{\sqrt{2r}} \sin\frac{\theta}{2}\cos\frac{\theta}{2}\cos\frac{3\theta}{2} + \cdots$$

$$t_{xy} = \frac{K_1}{\sqrt{2r}} \cos\frac{\theta}{2}\sin\frac{\theta}{2}\cos\frac{3\theta}{2} + \frac{K_2}{\sqrt{2r}} \cos\frac{\theta}{2}\left(1 - \sin\frac{\theta}{2}\sin\frac{3\theta}{2}\right) + \cdots$$

$$\begin{aligned} u_x = & \frac{K_1\sqrt{2r}}{8\mu}\left[(2\kappa - 1)\cos\frac{\theta}{2} - \cos\frac{3\theta}{2}\right] + \frac{K_2\sqrt{2r}}{8\mu}\left[(2\kappa + 1)\sin\frac{\theta}{2} + \sin\frac{3\theta}{2}\right] \\ & - \frac{\sigma - \chi}{8\mu}(\kappa + 1)(x + a) + \cdots \end{aligned}$$

$$\begin{aligned} u_y = & \frac{K_1\sqrt{2r}}{8\mu}\left[(2\kappa + 1)\sin\frac{\theta}{2} - \sin\frac{3\theta}{2}\right] - \frac{K_2\sqrt{2r}}{8\mu}\left[(2\kappa - 3)\cos\frac{\theta}{2} + \cos\frac{3\theta}{2}\right] \\ & - \frac{\sigma - \chi}{8\mu}(\kappa - 3)y + \cdots \end{aligned}$$

6. Prove analytically or numerically that the stresses satisfy the equilibrium equations.
7. ***Mini-Project.*** Consider the three-point bending beam with a small edge crack in section "Three-Point Bending Beam with a Small Edge Crack"; find the elastic material constants for standard steel.
   (a) Use the software for elastic static analysis to solve stress distribution.
   (b) Assume the applied loading is a Heaviside step function in time as shown in Fig. 6.14. Use the software for elastic dynamic analysis to solve the time history of the stress near crack tip.

7 Meshless Analysis of Nonlocal Continua

Introduction to Nonlocal Theory

Classical continuum mechanics is a local theory. It is based on the fundamental assumption that all balance laws are valid for every part of the body, however small it may be, and that the state of the body at any material point is influenced only by the state in the infinitesimal neighborhood about that point. The first of these assumptions eliminates the long-range effect of loads on the motion and the evolution of the state of the body, and the second ignores the effect of long-range interatomic interactions. This implies a certain intrinsic limitation (long-wavelength limit) since the cohesive forces in real materials have a finite or even infinite range, and nonlocality is an intrinsic aspect of certain material phenomena.

Generally, any problem that requires the solution of integrodifferential equations can be said to be nonlocal. Solid-state physics, and more particularly, lattice dynamics, is based on the dynamics of atoms attracted to each other by long-range forces. In the dislocation dynamics, the state of the body at a point is influenced appreciably by the dislocations that take place at neighboring points. Plastic deformations, known to originate at the atomic scale through the accumulations of large numbers of dislocations, are generally nonlocal in character. Suspensions, thin films, composites, bubbly fluids, and phase transitions are a few other fields where nonlocality may play an essential role.

Eringen (1966) proposed a nonlocal elastic theory in the spirit of the well-accepted classical continuum mechanics with certain modifications. Edelen (1969), Edelen and Laws (1971), and Edelen et al. (1971) proposed a nonlocal elastic theory in the tradition of the Gibbsian thermodynamics employing a variational principle. Both approaches turned out to give similar results for elastic solids (Eringen and Edelen, 1972). Later, further developments in this area are almost within the framework of those two theories.

The applications of nonlocal theory have shown the positive signs in explaining and predicting physical phenomena in microscopic length scales. The critical examples include Rayleigh surface waves with small wavelength, stress concentration at the tip of crack, and quasi-static elastic dielectrics. The dispersion curves obtained for both plane and Rayleigh surface waves fit perfectly with the atomic dispersion curves throughout the entire Brillouin zone, obtained by lattice dynamic computations. Moreover, the fracture toughness calculated agrees fairly well with experimental results on brittle materials. The recent application examples, including strain softening due to tunnel excavation, shear band formation, localization, and crack initiation and failure process, have also proved the success of nonlocal methods in eliminating the mesh dependence and in interpreting size effects.

The Framework of Nonlocal Theory

In continuum physics, nonlocal is taken to mean that the fundamental equations of a physical system contain integrals as well as derivatives of the dependent variables. For example, the basic equations of radiative transfer may be written as:

$$\mu \frac{\partial I}{\partial x}(x, \mu, \nu) + I(x, u, \nu) = \iint I(x, s, t) P(s, \mu, t, \nu)\, \mathrm{d}s\, \mathrm{d}t. \qquad (7.1)$$

This integrodifferential equation describes an intrinsically nonlocal process. Indeed, the disciplines of radiative transfer, particle interaction, stellar dynamics, and nuclear reactor theory are founded on integrodifferential equations. The mathematics of these disciplines is intimately connected with the nonlocal aspects of the problems and requires great care in order to ensure uniqueness and computational feasibility for solutions.

Time Nonlocality

Time nonlocality (memory effect) has been well recognized. A lot of experimental results available show that some materials have strong memory effects. In time nonlocal theory, the state of the body, at the material point **X** at time t, depends on the histories of motions and temperature of **X** at all times prior to and at t. The dependent constitutive variables are considered to be functionals of all the independent constitutive variables that are functions of all past times. For example, in Boltzmann–Volterra theory, the constitutive equation was expressed as

$$t_{ij}(t) = a_{ijmn} e_{mn}(t) + \int_{-\infty}^{t} b_{ijmn}(t - s) \frac{\partial e_{mn}(s)}{\partial s}\, \mathrm{d}s, \qquad (7.2)$$

where t_{ij} is the stress tensor and e_{mn} the strain. It is seen that the strain rates of all past times, s, are incorporated and the attenuation function **b** reflects that the axiom of memory is enforced. Therefore, rate-dependent theory accounts for certain time nonlocality.

Space Nonlocality

The simplest way of exemplifying the space nonlocal theories is to state that the dependent constitutive variables at a material point **x** are functionals of the independent constitutive variables at all points $\mathbf{x}'$ of the body, e.g.,

$$\mathbf{T}(\mathbf{X}, t) = \mathbf{F}\{\mathbf{E}(\mathbf{X}', t)\}, \quad \mathbf{X}' \in B. \tag{7.3}$$

This means the stress $\mathbf{T}$ at $\mathbf{X}$ and t is a functional of the strain $\mathbf{E}' \equiv \mathbf{E}(\mathbf{X}', t)$ that is a function of all the material points $\mathbf{X}'$ in body B and time t. The functional $\mathbf{F}$ in Eq. (7.3) may be approximated by a series of gradients of the strain tensor or by a series of multiple volume integrals.

In the approach proposed by Eringen, a difference function has been introduced

$$\mathbf{y}'(\mathbf{X}') \equiv \mathbf{y}(\mathbf{X} - \mathbf{X}') - \mathbf{y}(\mathbf{X}), \tag{7.4}$$

and with

$$\mathbf{y} = \mathbf{y}(\mathbf{X}, t), \tag{7.5}$$

$$\mathbf{y}' = \mathbf{y}'(\mathbf{X}', t), \tag{7.6}$$

to represent local and nonlocal (relative to $\mathbf{X}$) variables (e.g., strain, temperature, plastic strain, etc.), respectively, an asterisk placed on a quantity indicates the interchanges of $\mathbf{X}'$ and $\mathbf{X}$, e.g.,

$$\mathbf{A}^*(\mathbf{X}', \mathbf{X}, \mathbf{y}', \mathbf{y}) = \mathbf{A}(\mathbf{X}, \mathbf{X}', \mathbf{y}, \mathbf{y}'); \tag{7.7}$$

then the dependent variables $\mathbf{Z}$ (stress, heat flux, entropy, Helmholtz free energy, etc.) can be symbolically written as

$$\mathbf{Z}(\mathbf{X}, t) = \int_{V'} \mathbf{z}(\mathbf{y}, \mathbf{y}')\, \mathrm{d}V', \tag{7.8}$$

and the material time rate of Ψ can be obtained as

$$\begin{aligned} \dot{\Psi} &= \int_{V'} \left\{ \frac{\partial \psi}{\partial \mathbf{y}} \cdot \dot{\mathbf{y}} + \frac{\delta \psi}{\delta \mathbf{y}'} \cdot \dot{\mathbf{y}}' \right\} \mathrm{d}V' \\ &= \dot{\mathbf{y}} \cdot \int_{V'} \frac{\partial \psi}{\partial \mathbf{y}}\, \mathrm{d}V' + \int_{V'} \frac{\delta \psi}{\delta \mathbf{y}'} \cdot \dot{\mathbf{y}}'\, \mathrm{d}V' \\ &= \dot{\mathbf{y}} \cdot \frac{\partial \psi}{\partial \mathbf{y}} + \int_{V'} \frac{\delta \psi}{\delta \mathbf{y}'} \cdot \dot{\mathbf{y}}\, \mathrm{d}V', \end{aligned} \tag{7.9}$$

where the second term in the right-hand side of above equation involves the Frechet derivative $\delta\psi/\delta\mathbf{y}'$ (Griffel, 1981).

As far as the stress–strain relation is concerned, nonlocality means the stress at a point $\boldsymbol{x}$ is a function of the strains at all points of the body. For linear elasticity, this gives

$$t_{ij}(\mathbf{x}) = C_{ijkl} e_{kl}(\mathbf{x}) + \int c_{ijkl}(\mathbf{x}, \mathbf{x}') e_{kl}(\mathbf{x}')\, \mathrm{d}\mathbf{x}', \tag{7.10}$$

where $t_{ij}(\mathbf{x})$ is the stress at $\mathbf{x}$, $e_{ij}(\mathbf{x}')$ the strain at $\mathbf{x}'$, and C and c the material moduli tensors. The strain gradients would appear after the performance of the integrations when $e_{kl}(\mathbf{x}')$ is written as a Taylor series around point $\mathbf{x}$. This is why sometimes strain gradient theory is called nonlocal theory.

Material Instability and Intrinsic Length

Certain materials or processes exhibit narrow zones of intensive straining. This localization of deformation poses considerable difficulties in numerical solutions. From a mathematical point of view, it is due to a change of type of governing equations: loss of ellipticity in quasi-static problems and change from hyperbolic to elliptic type in the dynamic case. This change of type allows the prediction of the critical stress level that triggers localization; however, it leaves the size of the localization zone unspecified in static problems and gives infinite strains over a set of measure zero in dynamic problem. When incorporated into a computational model, strain-softening behavior therefore leads to severely mesh-dependent results in which deformation localizes in one element irrespective of its size. Furthermore, the energy dissipated in the strain-softening domain tends to diminish as the mesh is refined.

To remedy the change of type of governing equations, a length scale or timescale must be incorporated, implicitly or explicitly, into the material description or formulation of the boundary value problem. Rate-dependent material models account for certain time nonlocality and introduce a timescale. In strain gradient theories, the stress depends on strain and strain gradients. This accounts for some neighboring effects, i.e., the nonlocality in space, and thus introduces a length scale. As a consequence, the rate-dependent material does not lose strong ellipticity, and strain localization is caused by inhomogeneity; the strain gradient can be served as a regularization of material instabilities, and the mesh sensitivity can be remedied. However, there are at least two disadvantages in gradient-dependent approaches. First, corresponding to the strain gradients, there are high-order stresses; more material constants as well as additional boundary conditions are thus needed. Second, in the discretization, higher order differentiable shape functions are required, and the physical meaning of the additional boundary conditions is still an open issue.

The integral nonlocal theory naturally brings timescale or length scale into consideration. Its time or length effects are achieved by an attenuation function or a weight function that has clearly understood physical origin. It has an inherent advantage over either rate-dependent or gradient-dependent theories in numerical treatments.

A common feature of all meshless methods is a weight function that is defined to have compact support. The support size of the weight function is usually greater than the nodal spacing, and therefore nonlocality is embedded in the weight function and meshless method is nonlocal in nature. It is thus natural to lead to a conclusion that both the stability of meshless method and the numerical implementation of nonlocal field theory should benefit from the incorporation.

Nonlocal Constitutive Relations

In a nonlocal continuum, the state of the body at a material point $\mathbf{X}$ and time t cannot be determined entirely by the state variables at $\mathbf{X}$ and t. In this book, attention is focused on the nonlocality in the constitutive relations.

Generally, a nonlocal constitutive relation for viscoelastic solid takes the following form:

$$t_{ij}(\mathbf{x}) = \int_{\Omega'} A_{ijmn}(\mathbf{x}, \mathbf{x}') e_{mn}(\mathbf{x}')\, \mathrm{d}\Omega(\mathbf{x}') + \int_{\Omega'} a_{ijmn}(\mathbf{x}, \mathbf{x}') \dot{e}_{mn}(\mathbf{x}')\, \mathrm{d}\Omega(\mathbf{x}'). \tag{7.11}$$

For isotropic material, Eq. (7.11) is simplified to

$$t_{ij}(\mathbf{x}) = H(\mathbf{x}) \left\{ (\lambda \delta_{ij} \delta_{mn} + 2\mu \delta_{im} \delta_{jn}) \int_{\Omega'} f(r) e_{mn}(\mathbf{x}')\, \mathrm{d}\Omega(\mathbf{x}') + (\bar{\lambda} \delta_{ij} \delta_{mn} + 2\bar{\mu} \delta_{im} \delta_{jn}) \int_{\Omega'} f(r) \dot{e}_{mn}(\mathbf{x}')\, \mathrm{d}\Omega(\mathbf{x}') \right\}, \tag{7.12}$$

where $f(r)$ is the weight function,

$$f(r) = \begin{cases} 1 - 6r^2 + 8r^3 - 3r^4, & \text{if } r \leq 1 \\ 0, & \text{if } r > 1 \end{cases} \tag{7.13}$$

$$H(\mathbf{x}) \equiv \frac{1}{\int_{\Omega'} f(r)\, \mathrm{d}\Omega(\mathbf{x}')}, \tag{7.14}$$

where

$$r \equiv \frac{\|\mathbf{x} - \mathbf{x}'\|}{R},$$

and R is the radius of nonlocality. It is noted that the constitutive equation, Eq. (7.12), with Eqs. (7.13)–(7.14) has the following properties:

1. As R approaches zero, the stress tensor at $\mathbf{x}$ becomes

$$t_{ij}(\mathbf{x}) = A^0_{ijmn} e_{mn}(\mathbf{x}) + a^0_{ijmn} \dot{e}_{mn}(\mathbf{x}), \tag{7.15}$$

   which is the corresponding local constitutive equation for viscoelastic solid.
2. For constant strains and strain rates, the stresses are also constant.
3. For any interior point $\mathbf{x}$ whose distance from the nearest point on the boundary S is greater than R, the function $H(\mathbf{x})$ becomes a constant.

Also, it is emphasized that to apply the constitutive equation, Eq. (7.11) or (7.12), the visibility criterion has to be checked to determine whether the stress at $\mathbf{x}$ is influenced by the strain and strain rate at $\mathbf{x}'$.

Formulation of Nonlocal Meshless Method

Let the displacement field, $\mathbf{u}(\mathbf{x})$, be approximated as

$$u_i \cong \hat{u}_i = \Phi_{i\alpha} U_\alpha, \tag{7.16}$$

the strain field, $\mathbf{e}(\mathbf{x})$, can thus be obtained as

$$e_{ij}(\mathbf{x}) \cong \hat{e}_{ij}(\mathbf{x}) = \frac{1}{2}(\Phi_{i\alpha,j} + \Phi_{j\alpha,i}) U_\alpha = B_{ij\alpha}(\mathbf{x}) U_\alpha. \tag{7.17}$$

Following the procedures outlined in Chapter 3, the detailed expressions for the shape function Φ and its derivative $\Phi_{,i}$ can be easily found. Recall the strong form of continuum mechanics as

$$t_{ij,i} + \rho f_i - \rho \ddot{u}_j = 0 \quad \text{in } \Omega, \tag{7.18}$$

$$u_i = \bar{u}_i \quad \text{on } \Gamma_{ui}, \tag{7.19}$$

$$t_i \equiv t_{ki} n_k = \bar{t}_i \quad \text{on } \Gamma_{ti}, \tag{7.20}$$

where the union of the essential boundary, Γ_{ui}, and the natural boundary, Γ_{ti}, is the enclosing surface of the domain Ω, i.e., $\Gamma_{ti} \bigcup \Gamma_{ui} = \partial\Omega$ $(i = 1, 2, 3)$. If there is no ambiguity, let $\Gamma_u \equiv \bigcup_i \Gamma_{ui}$ and $\Gamma_t \equiv \bigcup_i \Gamma_{ti}$. One of the differences between the finite element (FE) methods and the meshless methods is that Eq. (7.16) is an approximation rather than an interpolation, i.e.,

$$\boldsymbol{\Phi}(\mathbf{x}_I) \cdot \mathbf{U} \neq u_{iI}, \tag{7.21}$$

where u_{iI} is the ith component of the displacement field at the Ith node, and therefore the essential boundary conditions, Eq. (7.19), should be read as

$$\Phi_{i\alpha} U_\alpha = \bar{u}_i \quad \text{on } \Gamma_{ui}, \tag{7.22}$$

which are constraints rather than specifications on $\mathbf{U}$.

The corresponding weak form based on the meshless particle methods can be obtained as

$$\int_\Omega \sigma_{ij} \delta e_{ij}\, dV + \int_\Omega \rho \ddot{u}_i \delta u_i\, dV - \int_{\Gamma_t} t_i \delta u_i\, dS - \int_\Omega \rho f_i \delta u_i\, dV$$
$$+ \int_{\Gamma_u} \lambda_i \delta u_i\, dS + \int_{\Gamma_u} \delta\lambda_i (u_i - \bar{u}_i)\, dS = 0, \tag{7.23}$$

where $\boldsymbol{\lambda}$ is the vector of Lagrange multiplies introduced here to enforce the essential boundary conditions, Eq. (7.19), on Γ_u.

Now, approximate $\boldsymbol{\lambda}$ on Γ_u in terms of nodal value $\boldsymbol{\Lambda}$ as

$$\lambda_i = \psi_{i\alpha} \Lambda_\alpha. \tag{7.24}$$

The weak form, Eq. (7.23), becomes

$$\delta U_\alpha \{ M_{\alpha\beta} \ddot{U}_\beta + C_{\alpha\beta} \dot{U}_\beta + K_{\alpha\beta} U_\beta - F_\alpha + G_{\alpha\beta} \Lambda_\beta \} + \delta \Lambda_\alpha \{ G_{\beta\alpha} U_\beta - f_\alpha \} = 0. \tag{7.25}$$

where

$$M_{\alpha\beta} = \int_\Omega \rho \Phi_{i\alpha} \Phi_{i\beta}\, d\Omega = M_{\beta\alpha}, \tag{7.26}$$

$$C_{\alpha\beta} = a^0_{ijmn} \int_\Omega \int_{\Omega'} H(\mathbf{x}) f(r) B_{ij\alpha}(\mathbf{x}) B_{mn\beta}(\mathbf{x}')\, d\Omega(\mathbf{x})\, d\Omega(\mathbf{x}'), \tag{7.27}$$

$$K_{\alpha\beta} = A^0_{ijmn} \int_\Omega \int_{\Omega'} H(\boldsymbol{x}) f(r) B_{ij\alpha}(\boldsymbol{x}) B_{mn\beta}(\boldsymbol{x}')\,\mathrm{d}\Omega(\boldsymbol{x})\,\mathrm{d}\Omega(\boldsymbol{x}'), \tag{7.28}$$

$$G_{\alpha\beta} = \int_{\Gamma_u} \Phi_{i\alpha} \Psi_{i\beta}\,\mathrm{d}S, \tag{7.29}$$

$$F_\alpha = \int_\Omega \rho f_i \Phi_{i\alpha}\,\mathrm{d}\Omega + \int_{\Gamma_t} \bar{t}_i \Phi_{i\alpha}\,\mathrm{d}S, \tag{7.30}$$

$$f_\alpha = \int_{\Gamma_u} \bar{u}_i \Psi_{i\alpha}\,\mathrm{d}S. \tag{7.31}$$

Because Eq. (7.25) should hold for any arbitrary $\delta\mathbf{U}$ and $\delta\mathbf{\Lambda}$, the governing equations in matrix form are thus obtained

$$\mathbf{M}\ddot{\mathbf{U}} + \mathbf{C}\dot{\mathbf{U}} + \mathbf{K}\mathbf{U} + \mathbf{G}\mathbf{\Lambda} = \mathbf{F}, \tag{7.32}$$

$$\mathbf{G}^t\mathbf{U} = \mathbf{f}. \tag{7.33}$$

For static problems, the governing equations are reduced to

$$\begin{vmatrix} \mathbf{K} & \mathbf{G} \\ \mathbf{G}^t & \mathbf{0} \end{vmatrix} \begin{vmatrix} \mathbf{U} \\ \mathbf{\Lambda} \end{vmatrix} = \begin{vmatrix} \mathbf{F} \\ \mathbf{f} \end{vmatrix}. \tag{7.34}$$

For dynamic problems, solve the following system of linear equations to get $\bar{\mathbf{K}}^t$ and $\bar{\mathbf{G}}^t$

$$\begin{vmatrix} \mathbf{K}^t & \mathbf{G} \\ \mathbf{G}^t & \mathbf{0} \end{vmatrix} \begin{vmatrix} \bar{\mathbf{K}}^t \\ \bar{\mathbf{G}}^t \end{vmatrix} = \begin{vmatrix} \mathbf{I} \\ \mathbf{0} \end{vmatrix}. \tag{7.35}$$

Then the governing equations for the displacements become

$$\bar{\mathbf{K}}\mathbf{M}\ddot{\mathbf{U}} + \bar{\mathbf{K}}\mathbf{C}\dot{\mathbf{U}} + \mathbf{U} = \bar{\mathbf{K}}\mathbf{F} + \bar{\mathbf{G}}\mathbf{f}. \tag{7.36}$$

These equations have the same form as local theory but with nonlocal constitutive relations. It is verified both analytically and numerically that the displacement field so obtained satisfies Eq. (7.19), the essential boundary conditions.

Numerical Examples by Nonlocal Meshless Method

Beam with Different Length Scales

A cantilever beam with fixed ratio of length/height, L/H, is modeled with uniformly distributed nodes, as shown in Fig. 7.1. The beam is subjected to an end load equivalent to an averaged shear stress $t_{xy} = t^0$. For the sake of comparison of results with different length scales, normalized deflection is defined as $\bar{u} \equiv u_y(x = L)E/Lt^0$ and length scale is defined as L/R, where R represents the size of the long-range interaction. The normalized deflection is plotted in length

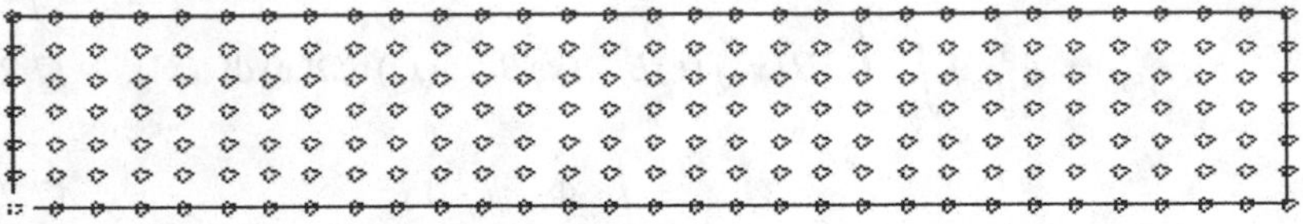

FIGURE 7.1. Meshless particle discretization of the beam.

scales for both meshless local and nonlocal theories in Fig. 7.2. It is seen that the solution of local theory is independent of the scales, which is well known. While in the nonlocal solution, the deflections are dependent of the length scales, it become larger and larger as the size decreases. This means that the material becomes less stiff and more compliant. This is consistent with the qualitative results obtained by atomistic simulations for silicon and quartz (Rudd and Broughton, 1999).

Static Analysis of a Cracked Specimen

Figure 7.3 is a cracked specimen. It has a width w and height $2h$. The horizontal edge crack extends from $x = 0$ to $x = a$. A uniformly distributed normal stress, $t_{yy} = \sigma(t)$, is applied along the top and bottom surfaces ($y = \pm h$). A plane strain condition is assumed.

The first case is a static case with $\sigma(t) = \sigma^0$. The h-refinement is implemented around the crack tip. The solutions of local and nonlocal ($R = 0.02w$) are obtained. The normalized stresses, $t_{yy}(x, y)/\sigma^0$, of these two cases together with the corresponding FE solution and the analytical crack tip solution (Sih and Liebowitz, 1968) are plotted as a function of normalized x coordination, x/w, along the line crack at $y/w = 0.005$ in Fig. 7.4.

In Fig. 7.4, the black line is the analytical crack tip solution that serves as a base reference to compare with the meshless numerical solutions. The green line is the

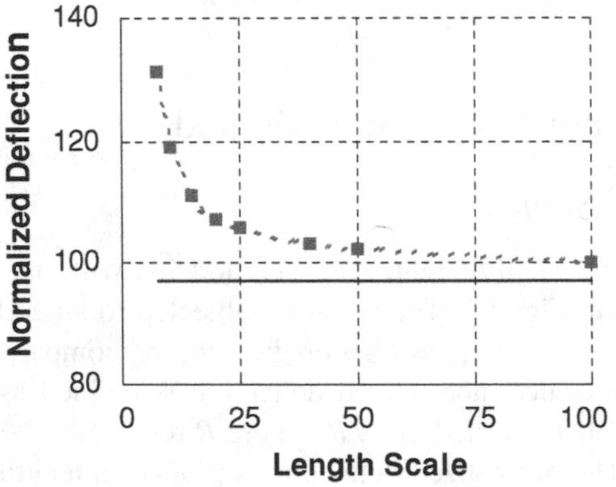

FIGURE 7.2. — Numerical and analytical solution of local theory; -■--■- numerical solution of nonlocal theory.

FIGURE 7.3. Edge cracked specimen subjected to mode-I tensile stress.

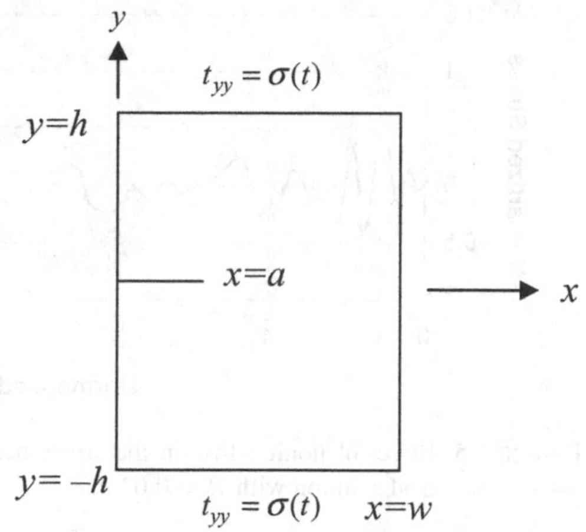

FE solution that is close to the analytical solution after the crack tip, but deviates off before the crack tip. The red line is the meshless solution of the local theory ($R = 0$). After the crack tip, it is qualitatively in agreement with the analytical solution, and before the crack tip, although it oscillates more than the FE solution, the unreasonably large stress at $x \to a^-$ disappears. With finer mesh for FE method and denser nodes for meshless method, the oscillation of the numerical solutions would disappear.

The common characteristics of the analytical, FE, and meshless solutions of local theory is the stress singularity. It is noted that the nonlocal meshless solution (the blue line) does not exhibit stress singularity at the crack tip; it yields a continuous stress field and has smaller oscillations before the crack tip; the stress distribution along the line crack closely resembles the relation between the cohesive force acting on bonds and the atomic positions (Masuda-Jindo et al., 1991).

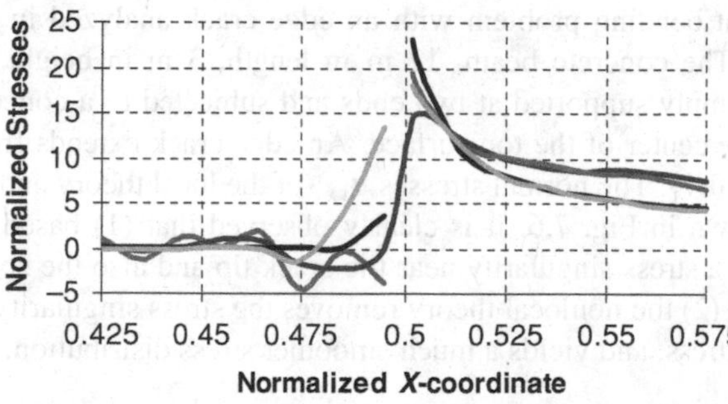

FIGURE 7.4. Stress distribution before and after the crack tip; red line: meshless solution of local theory; blue line: meshless solution of nonlocal theory with $R = 0.02w$; green line: finite element solution of local theory; black line: analytical crack tip solution of local theory.

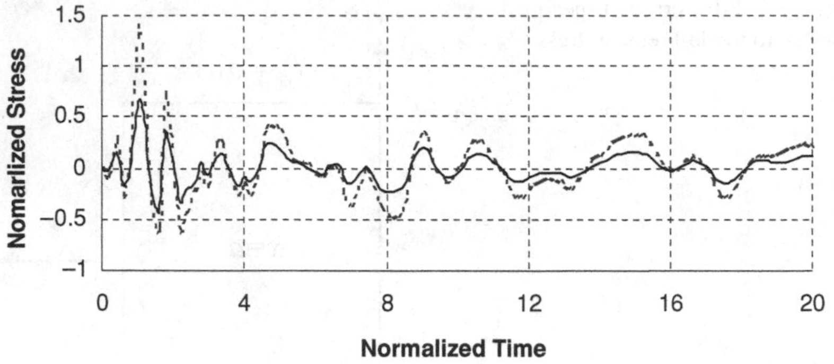

FIGURE 7.5. Effect of nonlocality on the stress near the crack tip. ------, local solution; ———, nonlocal solution with $R = 0.03w$.

Dynamic Analysis of the Cracked Specimen

The applied time-dependent loading in this case is expressed as

$$t_{yy}(x, \pm h, t) = \begin{cases} \sigma^0 \sin(2\pi t/t_0), & \text{if } t \leq t_0 \\ 0, & \text{if } t > t_0 \end{cases} \tag{7.37}$$

where $t_0 = 0.5\tau$ is small compared with the time needed for the longitudinal wave to reach the line crack from the edges where the loadings are applied. The normalized stress, t_{yy}/σ^0, at the sampling point nearest to the crack tip of two subcases, local solution with $R = 0$ and nonlocal solution with $R = 0.03w$, are plotted in Fig. 7.5. It is also seen that the magnitude of the stress in the nonlocal case is smaller than that in the local case. This is also qualitatively consistent with that by molecular dynamics simulation (Hoover, 1991).

Three-Point Bending Beam with a Small Edged Crack

The three-point bending problem with an edge crack analyzed in Chapter 6 is reconsidered. The concrete beam, 12 m in length, 3 m in height, and 1 m in thickness, is simply supported at two ends and subjected to a concentrated load of 100 N at the center of the top surface. An edge crack extends from $\{x, y\} = \{4, 0.0\}$ to $\{4, 0.4\}$. The normal stresses, t_{xx}, of the local theory and the nonlocal theory are shown in Fig. 7.6. It is clearly observed that (1) based on the local theory there is a stress singularity near the crack tip and also the solution has an oscillation and (2) the nonlocal theory removes the stress singularity, reduces the magnitude of stress, and yields a much smoother stress distribution.

Discussions

Nonlocal theory predicts that, because of material interaction, the material will become less rigid and the stresses near the crack tip become smaller as the length

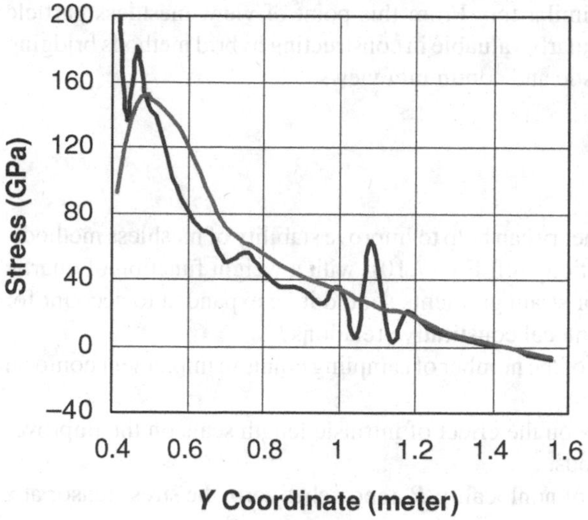

FIGURE 7.6. Stress t_{xx} distribution along the line crack (crack tip at $Y = 0.4$); blue line: meshless solution of local theory; red line: meshless solution of nonlocal theory.

scale decreases, in consistent with atomic model solutions. Most of all, the nonlocal solution eliminates the singularity at the crack tip and gives a continuous stress distribution. Nevertheless, it is conceptually and numerically very simple.

The nonlocal constitutive theory brings the influence of strains at distant point $\mathbf{x}'$ to the stresses at $\boldsymbol{x}$. Meshless particle methods share this essential feature by nonlocally constructing the approximation in the entire domain of influence. Meshless methods eliminate the need of constructing mesh and a lot of mesh-related problems and have great advantages in facilitating h and p adaptivity. The treatments of crack, shear band, and large deformation problems are the fields in which meshless methods have greatest promise. The crack, dislocation or plastic deformation, and small length scale problems such as MEMS, thin films, which are originally nonlocal in nature, are the fields in which nonlocality plays an essential role. Those are the promising fields where nonlocal meshless method has great application.

The stability of meshless methods is essential for their robustness. There are three instabilities for the meshless method (Belytschko et al., 2000): (1) the tensile instability, (2) the stability due to the spurious singular mode, and (3) the material instability. The material instability often occurs in the case of strain softening which results in a loss of ellipticity in the governing equation. It has been found that the particle equations with a Lagrangian kernel do not exhibit the tensile instability (Belytschko et al., 2000), and incorporating a length scale can remedy the loss of ellipticity (Chen and Belytschko, 2000). It is thus reasonable to expect that the stability of meshless methods can benefit from the incorporation with nonlocal constitutive theory.

The logical and computational structure of meshless methods closely resembles that of ordinary molecular dynamics, although additional state variables are required. The dynamics, the treatment of boundary conditions, and the analysis of chaotic instability are similar too. From this point of view, meshless particle method should prove particularly valuable in constructing hybrid methods bridging the gap between the atomistic and continuum views.

Problems

1. Explain why nonlocal theory can help to improve stability of meshless methods.
2. For nonlocal linear elasticity (cf. Eq. (7.10)) with a weight function of quartic spline, to which order of strain gradients should it be expanded to account for the effect of integral nonlocal constitutive relations?
3. What is the requirement of the number of sampling points to implement nonlocal constitutive relations?
4. Make a literature survey on the effect of intrinsic length scale on the improvement of meshless methods.
5. Prove that as the radius of nonlocality R approaches zero, the stress tensor at $\mathbf{x}$ becomes

$$t_{ij}(\mathbf{x}) = A^0_{ijmn} e_{mn}(\mathbf{x}) + a^0_{ijmn} \dot{e}_{mn}(\mathbf{x}),$$

   which is the corresponding local constitutive equation for viscoelastic solid. Hint: Let $f(r) \to \delta(r)$.
6. Prove that, from Eq. (7.12), if strains and strain rates are constant in space, then the stresses are also constant in space.
7. Show that for any interior point $\mathbf{x}$ whose distance from the nearest point on the boundary S is greater than R, the function $H(\mathbf{x})$ becomes a constant.

8
Meshless Analysis of Plasticity

In this chapter, we first present the formulation of a constitutive theory of thermoviscoelastoplastic (TVEP) continuum with full utilization of thermodynamics. No restrictive assumption has been made to the magnitude of any independent constitutive variables. Special attention is given to the notions of irreversibility and plastic dissipation. The fundamental concept of return mapping algorithm in computational plasticity is introduced. The special J_2 flow theory is presented in detail. The numerical procedures to obtain the solutions of boundary value problems by meshless method are outlined. As examples, the problems of finite strain high-speed impact and of slow crack growth in elastic–plastic solid are solved. Numerical results are presented and discussed.

Constitutive Relations

For the purpose of abbreviation, let the dependent constitutive variables be denoted by

$$\mathbf{Z} = \{\mathbf{T}, \mathbf{Q}, \psi, \eta\}, \tag{8.1}$$

and let the independent constitutive variables be decomposed into two sets

$$\mathbf{U} = \{\mathbf{E}, \theta\}, \tag{8.2}$$

$$\mathbf{V} = \{\dot{\mathbf{E}}, \dot{\theta}, \nabla\theta\}, \tag{8.3}$$

then, to construct the constitutive theory for thermoviscoelastic (TVE) material, one begins with (cf. Lee and Chen (2001) and Chapter 2)

$$\mathbf{Z} = \mathbf{Z}(\mathbf{U}, \mathbf{V}). \tag{8.4}$$

This seemingly general framework and approach elaborated in Chapter 2 turns out to be insufficient for the construction of the constitutive theory of plasticity. For plasticity, one needs to introduce a new set of variables, called internal variables, as

$$\mathbf{W} = \{\mathbf{E}^{\mathrm{p}}, \mathbf{R}\}, \tag{8.5}$$

where $\mathbf{E}^p$ is the plastic strain tensor corresponding to the Lagrangian strain $\mathbf{E}$, and $\mathbf{R}$, named as the hardening parameters, is a generalized vector of internal variables. Also, one may define the elastic strain tensor $\mathbf{E}^e$ as

$$\mathbf{E}^e \equiv \mathbf{E} - \mathbf{E}^p. \tag{8.6}$$

It should be emphasized that the (total) strain $\mathbf{E}$ is derivable from the displacement field; if the plastic strain $\mathbf{E}^p$ is obtained, then the elastic strain $\mathbf{E}^e$ follows the simple rule, Eq. (8.6).

To separate the material behavior into two distinct parts: TVE and TVEP, a scalar-valued yield function is expressed as

$$f = f(\mathbf{U}, \mathbf{V}, \mathbf{W}), \tag{8.7}$$

and, for a fixed set of values for $\mathbf{W}$, a hyper surface, named yield surface, is determined by the equation

$$f(\mathbf{U}, \mathbf{V}, \mathbf{W}) = 0. \tag{8.8}$$

The yield function can be chosen in such a way that the TVE region is corresponding to $f < 0$. The states of $f > 0$ are nonadmissible and ruled out in plasticity.

We now define the loading rate as the inner product between the outward normal to the yield surface and the tangent vector to the trajectory in the $\{\mathbf{U}, \mathbf{V}\}$ space, i.e.,

$$\begin{aligned} \xi &\equiv \frac{\partial f}{\partial \mathbf{U}} \cdot \dot{\mathbf{U}} + \frac{\partial f}{\partial \mathbf{V}} \cdot \dot{\mathbf{V}} \\ &= \frac{\partial f}{\partial \mathbf{E}} : \dot{\mathbf{E}} + \frac{\partial f}{\partial \theta}\dot{\theta} + \frac{\partial f}{\partial \dot{\mathbf{E}}} : \ddot{\mathbf{E}} + \frac{\partial f}{\partial \dot{\theta}}\ddot{\theta} + \frac{\partial f}{\partial \nabla\theta} \cdot \nabla\dot{\theta}. \end{aligned} \tag{8.9}$$

Three distinct cases, unloading, neutral loading, and loading, can be defined by (a) $f < 0$, (b) $f = \xi = 0$, and (c) $f = 0, \xi > 0$, respectively. The internal variables $\mathbf{W}$ will remain constant in cases of unloading and neutral loading. The evolution equations for the internal variables are postulated to be

$$\dot{\mathbf{W}} = \hat{\xi}\, \pi \phi(\mathbf{U}, \mathbf{V}, \mathbf{W}), \tag{8.10}$$

where $\hat{\xi} \equiv 0$ if $\xi \leq 0$ and $\hat{\xi} \equiv \xi$ if $\xi > 0$.

One should impose the consistency condition of plasticity: $f = 0$ and $\xi \geq 0$ lead to $\dot{f} = 0$, in other words, a TVEP state leads to another TVEP state. To enforce this consistency condition in case of loading, one must have

$$\begin{aligned} \dot{f} = 0 &= \frac{\partial f}{\partial \mathbf{U}} \cdot \dot{\mathbf{U}} + \frac{\partial f}{\partial \mathbf{V}} \cdot \dot{\mathbf{V}} + \frac{\partial f}{\partial \mathbf{W}} \cdot \xi\pi\varphi \\ &= \xi + \frac{\partial f}{\partial \mathbf{W}} \cdot \xi\pi\varphi, \end{aligned} \tag{8.11}$$

which implies

$$1 + \pi \frac{\partial f}{\partial \mathbf{W}} \cdot \varphi = 0. \tag{8.12}$$

Then π can be solved as

$$\pi = -\left\{\frac{\partial f}{\partial \mathbf{W}} \cdot \varphi\right\}^{-1}. \tag{8.13}$$

Now the Kuhn–Tucker conditions for plasticity can be written as

$$f \leq 0, \quad \xi \geq 0, \quad \xi f = 0. \tag{8.14}$$

Following the axiom of equipresence, the constitutive relations of TVEP material are now initiated as (also, take a note of Eqs. (8.7) and (8.10))

$$\mathbf{Z} = \mathbf{Z}(\mathbf{U}, \mathbf{V}, \mathbf{W}). \tag{8.15}$$

Substituting Eqs. (8.15) and (8.10) into the Clausius–Duhem inequality, Eq. (2.81), it results

$$-\rho^0\left\{\frac{\partial \psi}{\partial \mathbf{E}} \cdot \dot{\mathbf{E}} + \frac{\partial \psi}{\partial \theta}\dot{\theta} + \frac{\partial \psi}{\partial \dot{\mathbf{E}}} \cdot \ddot{\mathbf{E}} + \frac{\partial \psi}{\partial \dot{\theta}}\ddot{\theta} + \frac{\partial \psi}{\partial \nabla\theta} \cdot \nabla\dot{\theta} + \frac{\partial \psi}{\partial \mathbf{W}} \cdot \xi\pi\varphi + \eta\dot{\theta}\right\}$$
$$+\mathbf{T} \cdot \dot{\mathbf{E}} - \mathbf{Q} \cdot \nabla\theta/\theta \geq 0, \tag{8.16}$$

which implies

$$\psi = \psi(\mathbf{U}, \mathbf{W}) = \psi(\mathbf{E}, \theta, \mathbf{E}^{\mathrm{p}}, \mathbf{R}), \tag{8.17}$$

$$\eta = \eta^{\mathrm{e}} + \eta^d = -\frac{\partial \psi}{\partial \theta} + \eta^d(\mathbf{U}, \mathbf{V}, \mathbf{W}), \tag{8.18}$$

$$\mathbf{T} = \mathbf{T}^e + \mathbf{T}^d = \rho^0\frac{\partial \psi}{\partial \mathbf{E}} + \mathbf{T}^d(\mathbf{U}, \mathbf{V}, \mathbf{W}), \tag{8.19}$$

$$\mathbf{T}^d : \dot{\mathbf{E}} - \rho^0\eta^d\dot{\theta} - \mathbf{Q} \cdot \nabla\theta/\theta \geq 0, \tag{8.20}$$

$$\frac{\partial \psi}{\partial \mathbf{W}} \cdot \varphi \leq 0, \tag{8.21}$$

$$\frac{\partial f}{\partial \mathbf{W}} \cdot \varphi < 0, \tag{8.22}$$

where the last two inequalities came from the fact that π cannot be zero (cf. Eq. (1.12)); therefore it can never change sign, and without loss of generality, we choose $\pi > 0$.

In the following, we derive the governing equations of classical rate-independent plasticity in pure mechanical domain. First, we exclude $\mathbf{V} = \{\dot{\mathbf{E}}, \dot{\theta}, \nabla\theta\}$ from the list of independent constitutive variables. Then immediately we have

$$\eta = \eta^{\mathrm{e}} = -\frac{\partial \psi}{\partial \theta}, \quad \eta^d = 0, \tag{8.23}$$

$$\mathbf{T} = \mathbf{T}^{\mathrm{e}} = \rho^0\frac{\partial \psi}{\partial \mathbf{E}}, \quad \mathbf{T}^d = 0 \tag{8.24}$$

$$\mathbf{Q} = 0, \tag{8.25}$$

and inequality, Eq. (8.20), becomes a null statement. Green and Naghdi (1965), in their pioneer work on theory of plasticity, proposed that the stresses and entropy for thermoelastoplastic continuum are only functions of temperature and elastic Lagrangian strains. Following this idea, we further assume that the Helmholtz free energy density can be expressed as a polynomial of $\mathbf{E}^e = \mathbf{E} - \mathbf{E}^p$ and θ up to second order as

$$\begin{aligned} \psi &= \psi(\mathbf{E} - \mathbf{E}^p, \theta) = \psi(\mathbf{E}^e, \theta) \\ &= \left\{S^0 - \rho^0\eta^0 T - \tfrac{1}{2}\rho^0 c T^2/T^0 + \tfrac{1}{2}A_{KLMN}E^e_{KL}E^e_{MN}\right\}/\rho^0 \end{aligned} \tag{8.26}$$

then, from Eqs. (8.23) and (8.24), it results

$$\eta = \eta^0 + \frac{cT}{T^0}, \tag{8.27}$$

$$\mathbf{T} = \mathbf{A} : \mathbf{E}^e \text{ or } T_{KL} = A_{KLMN}E^e_{MN}, \tag{8.28}$$

where T^0 is the temperature of the natural state, may also be referred to as reference temperature, and $T \equiv \theta - T^0$ is the temperature variation from the reference temperature with the assumption that the temperature variation is small with respect to T^0, i.e.,

$$T^0 > 0, \quad |T| \ll T^0. \tag{8.29}$$

In such a way, the thermal and mechanical parts are separated and the second-order Piola–Kirchhoff (PK2) stress tensor is linearly proportional to the elastic Lagrangian strain tensor.

Now we may rewrite the yield function as

$$f = f(\mathbf{T}, \mathbf{R}). \tag{8.30}$$

Similarly, the evolution equations for the internal variables can be rewritten as

$$\dot{\mathbf{E}}^p = \gamma g(\mathbf{T}, \mathbf{R}), \tag{8.31}$$

$$\dot{\mathbf{R}} = -\gamma h(\mathbf{T}, \mathbf{R}), \tag{8.32}$$

where g and h are prescribed constitutive functions that define the direction of the plastic flow and type of hardening; γ is a nonnegative function and obeys the following Kuhn–Tucker conditions

$$\gamma \geq 0, \quad f(\mathbf{T}, \mathbf{R}) \leq 0, \quad \gamma f(\mathbf{T}, \mathbf{R}) = 0. \tag{8.33}$$

The flow rule, Eq. (8.31), and the hardening rule, Eq. (8.32), are for the general nonassociative model. For associative model, Eqs. (8.31) and (8.32) are specialized as

$$\dot{\mathbf{E}}^p = \gamma \frac{\partial f}{\partial \mathbf{T}}, \tag{8.34}$$

$$\dot{\mathbf{R}} = -\gamma \mathbf{D} \cdot \frac{\partial f}{\partial \mathbf{R}}, \tag{8.35}$$

where $\mathbf{D}$ is a symmetric matrix of plastic moduli and assumed to be positive definite. The physical meaning of Eq. (8.34) is that the plastic strain tensor is increasing in the direction parallel to the outward normal of the yield surface in the stress space.

Return Mapping Algorithm

In this section, we describe the numerical problems associated with the elastic–plastic process, which as we can see now is highly nonlinear, and a very powerful numerical algorithm, named *return mapping algorithm* (Simo and Hughes, 1998).

At time $t = t_n$, we assume that the total strain, the plastic strain, and all the other internal variables are known, i.e.,

$$\mathbf{E}(t_n) = \mathbf{E}_n, \quad \mathbf{E}^{\mathrm{p}}(t_n) = \mathbf{E}^{\mathrm{p}}_n, \quad \mathbf{R}(t_n) = \mathbf{R}^n. \tag{8.36}$$

The elastic strain and stress are dependent variables and can be calculated as

$$\mathbf{E}^{\mathrm{e}}(t_n) = \mathbf{E}^{\mathrm{e}}_n = \mathbf{E}_n - \mathbf{E}^{\mathrm{p}}_n, \tag{8.37}$$

$$\mathbf{T}(t_n) = \mathbf{T}_n = \mathbf{A} : \mathbf{E}^{\mathrm{e}}_n. \tag{8.38}$$

Now the problem is to update the basic variables $\mathbf{E}$, $\mathbf{E}^{\mathrm{p}}$, $\mathbf{R}$ at $t = t_{n+1}$ assuming that the incremental displacement field $\Delta\mathbf{U}$ is obtained through a finite element or a meshless analysis. The incremental displacement field being added to the displacement field at $t = t_n$ gives the displacement field $\mathbf{U}_{n+1}$ at $t = t_{n+1}$. Now this problem can be stated in the following mathematical form:

$$\mathbf{E}_{n+1} = \{\nabla_{\mathbf{X}}\mathbf{U}_{n+1} + (\nabla_{\mathbf{X}}\mathbf{U}_{n+1})^t + (\nabla_{\mathbf{X}}\mathbf{U}_{n+1}) \cdot (\nabla_{\mathbf{X}}\mathbf{U}_{n+1})^t\}/2, \tag{8.39}$$

$$\mathbf{T}_{n+1} = \mathbf{A} : (\mathbf{E}_{n+1} - \mathbf{E}^{\mathrm{p}}_{n+1}), \tag{8.40}$$

$$\mathbf{E}^{\mathrm{p}}_{n+1} = \mathbf{E}^{\mathrm{p}}_n + \Delta\gamma \frac{\partial f(\mathbf{T}_{n+1}, \mathbf{R}_{n+1})}{\partial \mathbf{T}}, \tag{8.41}$$

$$\mathbf{R}_{n+1} = \mathbf{R}_n - \Delta\gamma \mathbf{D} \cdot \frac{\partial f(\mathbf{T}_{n+1}, \mathbf{R}_{n+1})}{\partial \mathbf{R}}, \tag{8.42}$$

where $\Delta\gamma = \gamma_{n+1}\Delta t$, and the Kuhn–Tucker conditions become

$$f(\mathbf{T}_{n+1}, \mathbf{R}_{n+1}) \le 0, \quad \Delta\gamma \ge 0, \quad \Delta\gamma f(\mathbf{T}_{n+1}, \mathbf{R}_{n+1}) = 0. \tag{8.43}$$

The return mapping algorithm splits the problem into two parts, often referred as the elastic-predictor and the plastic-corrector problems. The elastic predictor creates a trial elastic state as follows:

$$\tilde{\mathbf{E}}^{\mathrm{p}}_{n+1} = \mathbf{E}^{\mathrm{p}}_n, \tag{8.44}$$

$$\tilde{\mathbf{R}}_{n+1} = \tilde{\mathbf{R}}_n, \tag{8.45}$$

$$\tilde{\mathbf{T}}_{n+1} = \mathbf{A} : (\mathbf{E}_{n+1} - \tilde{\mathbf{E}}^{\mathrm{p}}_{n+1}), \tag{8.46}$$

$$\tilde{f}_{n+1} = f(\tilde{\mathbf{T}}_{n+1}, \tilde{\mathbf{R}}_{n+1}), \tag{8.47}$$

where a "$\sim$" placed on a variable indicates the trial value of that variable at $t = t_{n+1}$.

An important concept in plasticity is the convexity of the elastic domain. A scalar-valued function g of a general vector $\mathbf{v}$ is said to be convex if, for $0 \leq \beta \leq 1$,

$$g(\beta \mathbf{v}_1 + (1-\beta)\mathbf{v}_2) \leq \beta g(\mathbf{v}_1) + (1-\beta) g(\mathbf{v}_2). \tag{8.48}$$

One may easily prove that $g(\mathbf{v})$ is convex if and only if

$$g(\mathbf{v}_1) - g(\mathbf{v}_2) \geq (\mathbf{v}_1 - \mathbf{v}_2) \cdot \frac{\partial g}{\partial \mathbf{v}}. \tag{8.49}$$

The convexity of the elastic domain, equivalent to that the yield function $f(\mathbf{T}, \mathbf{R})$ is convex. Simo and Hughes (1998) proved that

$$\tilde{f}_{n+1} - f_{n+1} \geq \Delta\gamma \left\{ \frac{\partial f_{n+1}}{\partial \mathbf{T}} : \mathbf{A} : \frac{\partial f_{n+1}}{\partial \mathbf{T}} + \frac{\partial f_{n+1}}{\partial \mathbf{R}} : \mathbf{D} : \frac{\partial f_{n+1}}{\partial \mathbf{R}} \right\}. \tag{8.50}$$

Since $\mathbf{A}$ and $\mathbf{D}$ are positive definite, Simo and Hughes (1988) further show that

$$\tilde{f}_{n+1} \leq 0 \;\Rightarrow\; f_{n+1} \leq 0 \quad \text{and} \quad \Delta\gamma = 0, \tag{8.51}$$

$$\tilde{f}_{n+1} > 0 \;\Rightarrow\; f_{n+1} = 0 \quad \text{and} \quad \Delta\gamma > 0. \tag{8.52}$$

This means the possible situation of loading/neutral or loading/unloading is solely decided by $\tilde{f}_{n+1}$. For the case $\tilde{f}_{n+1} > 0$, the Kuhn–Tucker conditions are violated by the trial elastic state and the consistency has to be restored by the plastic corrector. Dividing Eqs. (8.41) and (8.42) by $\Delta\gamma$, the plastic-corrector problem can be rephrased as a nonlinear initial value problem governed by the following ordinary differential equations and initial conditions for $\mathbf{E}^{\mathrm{p}}(\Delta\gamma)$ and $\mathbf{R}(\Delta\gamma)$:

$$\frac{d\mathbf{E}^{\mathrm{p}}}{d\delta\gamma} = \frac{\partial f(\mathbf{A} : (\mathbf{E}_{n+1} - \mathbf{E}^{\mathrm{p}}), \mathbf{R})}{\partial \mathbf{T}}, \tag{8.53}$$

$$\frac{\mathrm{d}\mathbf{R}}{\mathrm{d}\Delta\gamma} = -\mathbf{D} \cdot \frac{\partial f(\mathbf{A} : (\mathbf{E}_{n+1} - \mathbf{E}^{\mathrm{p}}), \mathbf{R})}{\partial \mathbf{R}}, \tag{8.54}$$

$$\mathbf{E}^{\mathrm{p}}(\Delta\gamma = 0) = \mathbf{E}^{\mathrm{p}}_n, \tag{8.55}$$

$$\mathbf{R}(\Delta\gamma = 0) = \mathbf{R}_n. \tag{8.56}$$

This means the internal variables are the solutions of the initial value problems specified by the differential equations, Eqs. (8.53) and (8.54), and the initial conditions, Eqs. (8.55) and (8.56). Equations (8.53)–(8.56) can also be expressed in stress space as

$$\frac{\mathrm{d}\mathbf{T}}{\mathrm{d}\Delta\gamma} = -\mathbf{A} : \frac{\partial f(\mathbf{T}, \mathbf{R})}{\partial \mathbf{T}}, \tag{8.57}$$

$$\frac{\mathrm{d}\mathbf{R}}{\mathrm{d}\Delta\gamma} = -\mathbf{D} \cdot \frac{\partial f(\mathbf{T}, \mathbf{R})}{\partial \mathbf{R}}, \tag{8.58}$$

$$\mathbf{T}(\Delta\gamma = 0)x = \tilde{\mathbf{T}}_{n+1}, \tag{8.59}$$

$$\mathbf{R}(\Delta\gamma = 0) = \tilde{\mathbf{R}}_{n+1} = \mathbf{R}_n. \tag{8.60}$$

In other words, we need to solve Eqs. (8.57) and (8.58) to obtain **T** and **R** as function of $\Delta\gamma$ starting from the trial elastic state, Eqs. (8.59) and (8.60), until consistency is restored by returning to the boundary of the elastic domain, equivalently, by the value of $\Delta\gamma$ such that

$$f(\mathbf{T}(\Delta\gamma), \mathbf{R}(\Delta\gamma)) = 0. \tag{8.61}$$

J_2 Flow Theory

In this section, we introduce a special and classical model for metal plasticity, often named as the J_2 flow theory. It has two distinct forms and needs different treatments, one for three-dimensional and plane strain problems and another for plane stress problems (Simo and Hughes, 1998).

J_2 Flow Theory for Three-Dimensional and Plane Strain Problems

First, we define the deviatoric stress tensor, with reference to the PK2 stress tensor, as

$$S_{KL} \equiv T_{KL} - \frac{1}{3}T_{MM}\delta_{KL} \text{ or } \mathbf{S} \equiv \mathbf{T} - \frac{1}{3}\text{tr}(\mathbf{T})\mathbf{I}. \tag{8.62}$$

Second, we assume that the material in question is isotropic, and then the fourth-order elastic property tensor **A** can be reduced to

$$A_{KLMN} = \lambda\delta_{KL}\delta_{MN} + \mu(\delta_{KM}\delta_{LN} + \delta_{KN}\delta_{LM}), \tag{8.63}$$

where λ and μ are the two Lame constants.

It is worthwhile at this moment to recall Eqs. (2.77) and (2.78)

$$T_{KL} = jt_{kl}X_{K,k}X_{L,l}, \quad t_{kl} = j^{-1}T_{KL}x_{k,K}x_{l,L}. \tag{8.64}$$

Note that, in small strain theory, one has the approximations: $x_{k,K} \cong \delta_{kK}$, $X_{K,k} \cong \delta_{Kk}$, and $\rho \cong \rho^0 \Rightarrow j \cong 1$, which make Eq. (8.64) read as

$$\begin{aligned} T_{KL} &= jt_{kl}X_{K,k}X_{L,l} \cong t_{kl}\delta_{Kk}\delta_{Ll} \\ t_{kl} &= j^{-1}T_{KL}x_{k,K}x_{l,L} \cong T_{KL}\delta_{kK}\delta_{lL}. \end{aligned} \tag{8.65}$$

In other words, in small strain theory, there is no difference between the Cauchy stress tensor and the second-order Piola–Kirchhoff stress tensor. What we derive in the following is valid for finite strain theory and certainly it is applicable to small strain theory.

From Eq. (8.62), it is seen that the trace of the deviatoric stress tensor is vanishing, i.e., $\text{tr}(\mathbf{S}) = S_{KK} \equiv 0$. A choice of internal variables, typical for metal plasticity, is

$$\mathrm{R} = \{\boldsymbol{\beta}, \bar{\beta}\}, \tag{8.66}$$

where $\boldsymbol{\beta}$ is a second-order symmetric tensor with $\text{tr}(\boldsymbol{\beta}) = 0$ and defines the center of the von Mises yield surface in the stress space and $\bar{\beta}$ defines the isotropic hardening of the von Mises yield surface. The resulting J_2 model has the following yield condition, flow rule and hardening law:

$$f(\mathbf{T}, \mathbf{R}) = ||\boldsymbol{\xi}|| - \sqrt{2/3}(\sigma_Y + cH\bar{\beta}), \tag{8.67}$$

$$\dot{\mathbf{E}}^{\text{p}} = \gamma \frac{\boldsymbol{\xi}}{||\boldsymbol{\xi}||}, \tag{8.68}$$

$$\dot{\boldsymbol{\beta}} = \gamma \frac{2}{3}(1-c)H \frac{\boldsymbol{\xi}}{||\boldsymbol{\xi}||}, \tag{8.69}$$

$$\dot{\bar{\beta}} = \gamma\sqrt{2/3}, \tag{8.70}$$

where

$$\boldsymbol{\xi} \equiv \mathbf{S} - \boldsymbol{\beta}, \quad ||\boldsymbol{\xi}|| \equiv \sqrt{\boldsymbol{\xi} : \boldsymbol{\xi}}, \tag{8.71}$$

σ_Y is the von Mises strength; H is a constant representing the slope of the stress–strain relation in the plastic loading; and c is a constant with $0 \leq c \leq 1$. Since, from Eq. (8.69), $||\dot{\mathbf{E}}^{\text{p}}|| = \gamma$, it is seen that Eq. (8.70) implies

$$\bar{\beta}(t) = \int_0^t \sqrt{2/3}||\dot{\mathbf{E}}^{\text{p}}(\tau)||d\tau, \tag{8.72}$$

which is the reason that $\bar{\beta}$ is usually named as the equivalent plastic strain. Also, it is noted that

1. $c = 1$implies $\dot{\boldsymbol{\beta}} = 0$ which is the case of isotropic hardening;
2. $c = 0$ implies that $\bar{\beta}$ does not affect the size of the yield surface—it is the case of kinematic hardening.

For this special J_2 model, one may obtain an exact closed-form solution for $\Delta\gamma$ in using the return mapping algorithm:

$$\tilde{f}_{n+1} = ||\tilde{\mathbf{S}}_{n+1} - \boldsymbol{\beta}_n|| - \sqrt{2/3}(\sigma_Y + cH\bar{\beta}_n), \tag{8.73}$$

$$\Delta\gamma = \frac{\tilde{f}_{n+1}}{2\mu + (2/3)H}. \tag{8.74}$$

Then the plastic strain and internal variable at $t = t_{n+1}$ can be updated as

$$\mathbf{E}^{\text{p}}_{n+1} = \mathbf{E}^{\text{p}}_n + \Delta\gamma \frac{\tilde{\mathbf{S}}_{n+1} - \boldsymbol{\beta}_n}{||\tilde{\mathbf{S}}_{n+1} - \boldsymbol{\beta}_n||}, \tag{8.75}$$

$$\boldsymbol{\beta}_{n+1} = \boldsymbol{\beta}_n + \frac{2}{3}(1-c)H\Delta\gamma \frac{\tilde{\mathbf{S}}_{n+1} - \boldsymbol{\beta}_n}{||\tilde{\mathbf{S}}_{n+1} - \boldsymbol{\beta}_n||}, \tag{8.76}$$

$$\bar{\beta}_{n+1} = \bar{\beta}_n + \sqrt{2/3}\Delta\gamma. \tag{8.77}$$

One may readily prove that with the updated values for $\mathbf{E}^{\text{p}}_{n+1}, \boldsymbol{\beta}_{n+1}, \bar{\beta}_{n+1}$ from Eqs. (8.75) to (8.77) and the updated values of $\mathbf{E}^{\text{e}}_{n+1}, \mathbf{T}_{n+1}, \mathbf{S}_{n+1}$ the value of the yield function

$$f_{n+1} = ||\mathbf{S}_{n+1} - \boldsymbol{\beta}_{n+1}|| - \sqrt{2/3}(\sigma_Y + cH\bar{\beta}_{n+1})$$

indeed returns to zero.

J_2 Flow Theory for Plane Stress Problems

Since, in the case of plane stress, $T_{31} = T_{32} = T_{33} = 0$, it is convenient to define the following.

$$\boldsymbol{\sigma} \equiv \{T_{11}, T_{22}, T_{12}\}^t, \tag{8.78}$$

$$\boldsymbol{\varepsilon} \equiv \{\varepsilon_{11}, \varepsilon_{22}, 2\varepsilon_{12}\}^t, \tag{8.79}$$

$$\boldsymbol{\varepsilon}^{\mathrm{p}} \equiv \left\{\varepsilon^{\mathrm{p}}_{11}, \varepsilon^{\mathrm{p}}_{22}, 2\varepsilon^{\mathrm{p}}_{12}\right\}^t, \tag{8.80}$$

$$\mathbf{P} \equiv \frac{1}{3}\begin{vmatrix} 2 & -1 & 0 \\ -1 & 2 & 0 \\ 0 & 0 & 6 \end{vmatrix}, \tag{8.81}$$

$$\mathbf{C} \equiv \frac{E}{1-v^2}\begin{vmatrix} 1 & v & 0 \\ v & 1 & 0 \\ 0 & 0 & \dfrac{1-v}{2} \end{vmatrix}, \tag{8.82}$$

where E and v are Young's modulus and Poisson's ratio.

A special J_2 model for plane stress problem may be represented by

$$\boldsymbol{\xi} \equiv \boldsymbol{\sigma} - \boldsymbol{\beta}, \tag{8.83}$$

$$f = (\boldsymbol{\xi} \cdot \mathbf{P} \cdot \boldsymbol{\xi})^{1/2} - \sqrt{2/3}K(\bar{\beta}), \tag{8.84}$$

$$\dot{\boldsymbol{\varepsilon}}^{\mathrm{p}} = \gamma \mathbf{P} \cdot \boldsymbol{\xi}, \tag{8.85}$$

$$\dot{\boldsymbol{\beta}} = \gamma \frac{2}{3} H \boldsymbol{\xi}, \tag{8.86}$$

$$\dot{\bar{\beta}} = \gamma \sqrt{2/3}(\boldsymbol{\xi} \cdot \mathbf{P} \cdot \boldsymbol{\xi})^{1/2}, \tag{8.87}$$

$$\boldsymbol{\sigma} = \mathbf{C} \cdot (\boldsymbol{\varepsilon} - \boldsymbol{\varepsilon}^{\mathrm{p}}) \tag{8.88}$$

The corresponding return mapping algorithm stipulates

$$\boldsymbol{\varepsilon}^{\mathrm{p}}_{n+1} = \boldsymbol{\varepsilon}^{\mathrm{p}}_{n} + \Delta\gamma \mathbf{P} \cdot \boldsymbol{\xi}_{n+1}, \tag{8.89}$$

$$\boldsymbol{\beta}_{n+1} = \boldsymbol{\beta}_n + \frac{2}{3}\Delta\gamma H \boldsymbol{\xi}_{n+1}, \tag{8.90}$$

$$\bar{\beta}_{n+1} = \bar{\beta}_n + \sqrt{2/3}\Delta\gamma \bar{f}, \tag{8.91}$$

where

$$\bar{f} \equiv (\boldsymbol{\xi}_{n+1} \cdot \mathbf{P} \cdot \boldsymbol{\xi}_{n+1})^{1/2}, \tag{8.92}$$

and $\Delta\gamma$ is the solution of the following nonlinear scalar equation

$$\bar{f}^2 = \frac{2}{3}\left[K(\bar{\beta}_n + \sqrt{2/3}\Delta\gamma \bar{f})\right]^2 \tag{8.93}$$

Simo et al. (1988) showed that, for linear kinematic hardening and certain forms of isotropic hardening, the discrete consistency equation reduces to a quartic equation that can be solved in closed form. For the case of pure kinematic hardening, the shape of the yield surface does not change, i.e., Eq. (8.93) can be rewritten as

$$\bar{f}^2(\Delta\gamma) = \frac{2}{3}\sigma_Y^2 \triangleq R^2, \tag{8.94}$$

which leads to

$$\begin{aligned}\bar{\gamma}^4 &+ \left(\frac{2}{C_3} + \frac{2}{C_4}\right)\bar{\gamma}^3 + \left(\frac{4}{C_3 C_4} + \frac{1 - C_2}{C_4^2} + \frac{1 - C_1}{C_3^2}\right)\bar{\gamma}^2 \\ &+ 2\left(\frac{1 - C_2}{C_3 C_4^2} + \frac{1 - C_1}{C_4 C_3^2}\right)\bar{\gamma} + \frac{1 - C_1 - C_2}{C_3^2 C_4^2} = 0,\end{aligned} \tag{8.95}$$

where

$$C_1 = \frac{(\tilde{\xi}_{11} + \tilde{\xi}_{22})^2}{6R^2}, \tag{8.96}$$

$$C_2 = \frac{(\tilde{\xi}_{11} - \tilde{\xi}_{22})^2 + 4\tilde{\xi}_{12}^2}{2R^2}, \tag{8.97}$$

$$C_3 = \frac{1}{3(1 - \upsilon)} + \frac{2H}{3E}, \tag{8.98}$$

$$C_4 = \frac{1}{1 + \upsilon} + \frac{2H}{3E}, \tag{8.99}$$

and $\tilde{\boldsymbol{\xi}}$ is the trial value of $\boldsymbol{\xi}$ at $t = t_{n+1}$; $\bar{\gamma} \equiv E\Delta\gamma$.

It can be further showed that this fourth-order algebraic equation has one positive root, one negative root, and a pair of complex conjugate roots (Hungerford, 1974; Herstein, 1964). In other words, $\Delta\gamma$ can be analytically and uniquely determined.

Meshless Analysis of High-Speed Impact/Contact Problem

The underlying structure of the finite element method that originates from their reliance on a mesh is not well suited to the treatment of extreme mesh distortion. As a consequence, the application of finite element method to large deformation problem involving severe mesh distortion is limited. On the other hand, meshless methods require no explicit mesh in computation (Belytschko et al., 1994, 1996, 2000a; Chen et al., 2002; Jun et al., 1998) and therefore avoid mesh distortion difficulties in large deformation analysis. One of the application areas in which meshless has obvious advantages over finite element method is high-speed impact problem.

High-speed impact is a dynamic large-strain elastic–plastic problem. Central difference method is utilized to find solutions as functions of time. It should also be noted that high-speed impact is a contact problem between two material bodies. During the process of contact, penetration is prohibited but separation is allowed. In fact, separation begins as soon as the pair of points in contact has the tendency to move apart. In meshless method, the displacement, as well as the velocity and the acceleration, of a point is not equal to the nodal value associated with that point; rather, it is a combination of nodal values of points in the neighborhood of that point; however, the instantaneous contact force being incorporated in the central difference method is only applicable to the point in question. This makes meshless analysis of contact problem very challenging and requires special attention during the formulation as well as the numerical implementation.

We recall the momentum equation (balance of linear momentum) as

$$t_{ji,j} + \rho f_i - \rho \dot{v}_i = 0. \tag{8.100}$$

Take the inner product between the momentum equation and a test function (here we choose the test function to be the virtual velocity) and integrate it over the current (deformed) configuration, it results

$$\int_v \delta v_i (t_{ji,j} + \rho f_i - \rho \dot{v}_i)\mathrm{d}v = 0. \tag{8.101}$$

Assuming that (1) virtual velocity vanishes on all the boundary on which essential boundary conditions are specified and (2) stresses are continuous on all material interfaces (this requirement is easily guaranteed by using meshless method), Eq. (8.101) leads to

$$\int_v \delta v_i \rho \dot{v}_i \mathrm{d}v = \int_v \delta v_i \rho f_i \mathrm{d}v + \int_{\Gamma_t} \delta v_i \bar{t}_i \mathrm{d}s - \int_v t_{ji} \delta v_{i,j} \mathrm{d}v, \tag{8.102}$$

where the second term in the RHS is the surface integral over all the boundaries on which the surface traction is specified as (with $\mathbf{n}$ being the unit outward normal on Γ_t)

$$n_j t_{ji} \equiv t_i = \bar{t}_i. \tag{8.103}$$

In meshless method, as well as in finite element method, let $N_\alpha(\mathbf{X})$ be the shape functions where $\mathbf{X}$ denotes the Lagrangian coordinate and α is referring to the αth node. Then the unknown function (trial function), in this case the displacement vector, and the test function at any generic point $\mathbf{X}$ can be expressed as

$$\begin{aligned} u_i(\mathbf{X}, t) &= N_\alpha(\mathbf{X}) U_{i\alpha}(t), \\ \delta v_i(\mathbf{X}, t) &= N_\alpha(\mathbf{X}) \delta V_{i\alpha}(t), \end{aligned} \tag{8.104}$$

where $U_{i\alpha}$ and $\delta V_{i\alpha}$ are the ith component of nodal values of the displacement and virtual velocity of the αth node, respectively. Then those terms in Eq. (8.102) can be further derived as

$$\begin{aligned}\int_v \delta v_i \rho f_i \mathrm{d}v + \int_{\Gamma_t} \delta v_i \bar{t}_i \mathrm{d}s &= \delta V_{i\alpha} \left\{ \int_v N_\alpha \rho f_i \mathrm{d}v + \int_{\Gamma_t} N_\alpha \bar{t}_i \mathrm{d}s \right\} \\ &= \delta V_{i\alpha} \left\{ \int_V N_\alpha \rho^0 f_i \mathrm{d}V + \int_{\Gamma_t^0} N_\alpha \bar{t}_i^0 \mathrm{d}S \right\} \\ &\equiv \delta V_{i\alpha} f_{i\alpha}^{\mathrm{ext}},\end{aligned} \tag{8.105}$$

$$\begin{aligned}\int_v t_{ji} \delta v_{i,j} \mathrm{d}v &= \delta V_{i\alpha} \int_v t_{ji} N_{\alpha,j} \mathrm{d}v \\ &= \delta V_{i\alpha} \int_v j^{-1} x_{i,K} x_{j,L} T_{KL} N_{\alpha,j} \mathrm{d}v \\ &= \delta V_{i\alpha} \int_V x_{i,K} T_{KL} N_{\alpha,L} \mathrm{d}V \\ &= \delta V_{i\alpha} \int_V (\delta_{iK} + N_{\beta,K} U_{i\beta}) T_{KL} N_{\alpha,L} \mathrm{d}V \\ &\equiv \delta V_{i\alpha} f_{i\alpha}^{\mathrm{int}},\end{aligned} \tag{8.106}$$

$$\int_v \delta v_i \rho \dot{v}_i \mathrm{d}v = \delta V_{i\alpha} \left\{ \int_v N_\alpha N_\beta \rho \mathrm{d}v \right\} \dot{V}_{i\beta} = \delta V_{i\alpha} \left\{ \int_V N_\alpha N_\beta \rho^0 \mathrm{d}V \right\} \dot{V}_{i\beta}. \tag{8.107}$$

Because Eq. (8.102) has to stand for arbitrary $\delta V_{i\alpha}$, we then derive

$$M_{\alpha\beta} \dot{V}_{i\beta} = f_{i\alpha}^{\mathrm{ext}} - f_{i\alpha}^{\mathrm{int}}, \tag{8.108}$$

where the mass matrix, defined as

$$M_{\alpha\beta} \equiv \int_V \rho^0 N_\alpha N_\beta \mathrm{d}V, \tag{8.109}$$

is constant in time and can be calculated in the Lagrangian setting. It should be emphasized that the formulation (including governing equation, Eq. (8.108), and definitions for $f_{i\alpha}^{\mathrm{ext}}$, $f_{i\alpha}^{\mathrm{int}}$, $M_{\alpha\beta}$ in Eqs. (8.105), (8.106), and (8.109), respectively) is exact for dynamic finite strain plasticity. Now we diagonalize the mass matrix by row-sum technique

$$M_{\underline{\alpha}\,\underline{\alpha}}^d = \sum_\beta M_{\alpha\beta}, \tag{8.110}$$

where M^d is the diagonal, lumped mass matrix and the bars underlying the α indicate that usually understood summation convention is suspended. Now Eq. (8.108) is reduced to

$$\dot{V}_{i\alpha} = \left(f_{i\underline{\alpha}}^{\mathrm{ext}} - f_{i\underline{\alpha}}^{\mathrm{int}} \right) \Big/ M_{\underline{\alpha}\,\underline{\alpha}}^d. \tag{8.111}$$

For abbreviation, we now rewrite Eq. (8.111) as

$$\mathbf{a}^n \equiv \ddot{\mathbf{u}}^n = \mathbf{f}^n / M^d, \tag{8.112}$$

where the superscript n is referring to solution at $t = t^n$. The procedures of solving the governing equation, Eq. (8.112), by the central difference method are given as follows (Belytschko et al., 2000b):

Step 1. Partially update the nodal velocities.

$$\mathbf{v}^{n+1/2} = \mathbf{v}^n + \frac{1}{2}\Delta t\mathbf{a}^n. \tag{8.113}$$

(At this moment, we assume that we know $\mathbf{u}^n$, $\mathbf{v}^n$, $\mathbf{a}^n$. This assumption is perfectly all right because at $t = 0$ we know $\mathbf{u}^0$ and $\mathbf{v}^0$ and from Eq. (8.112) we know $\mathbf{a}^0$.)

Step 2. Enforce velocity boundary conditions. Set

$$v_i\left(\mathbf{X}_\gamma, t^{(n+1)/2}\right) = \bar{v}_i\left(\mathbf{X}_\gamma, t^{(n+1)/2}\right), \tag{8.114}$$

if the γth node is one of those, we specify the ith component of the velocity at $t = t^{(n+1)/2}$ to be $\bar{v}_i\left(\mathbf{X}_\gamma, t^{(n+1)/2}\right)$. Note that in finite element analysis, Eq. (8.114) simply reads as

$$V_{i\gamma}^{(n+1)/2} = \bar{v}_i(\mathbf{X}_\gamma, t^{(n+1)/2}). \tag{8.115}$$

However, in meshless method, we know that

$$N_\alpha(\mathbf{X}_\beta) \neq \delta_{\alpha\beta}, \tag{8.116}$$

$$u_i(\mathbf{X}_\alpha, t) \neq U_{i\alpha}(t), \quad v_i(\mathbf{X}_\alpha, t) \neq V_{i\alpha}(t). \tag{8.117}$$

The velocity boundary conditions, Eq. (8.114), should be enforced by finding $V_{i\gamma}^{(n+1)/2}$ such that the following is satisfied

$$V_{i\beta}^{(n+1)/2} N_\beta(\mathbf{X}_\gamma) = \bar{v}_i(\mathbf{X}_\gamma, t^{1/2}). \tag{8.118}$$

This step should be performed with extreme caution.

Step 3. Update nodal displacements.

$$\mathbf{u}^{n+1} = \mathbf{u}^n + \Delta t\, \mathbf{v}^{(n+1)/2}. \tag{8.119}$$

Step 4. Computer nodal force using Eqs. (8.105) and (8.106).

Step 5. Compute $\mathbf{a}^{n+1}$ using Eq. (8.112).

Step 6. Partially update nodal velocities.

$$\mathbf{v}^{n+1} = \mathbf{v}^{(n+1)/2} + \frac{1}{2}\Delta t\mathbf{a}^{n+1}. \tag{8.120}$$

Step 7. Update counter, let $n \leftarrow n+1$. Go to Step 1 if the entire process of simulation has not been completed yet.

Note that the central difference method is an explicit method of which the condition of stability requires that the time step $\Delta t \equiv t^{n+1} - t^n$ cannot exceed a critical value Δt_c, which is given by

$$\Delta t_c = \frac{2}{\omega_{\max}}, \tag{8.121}$$

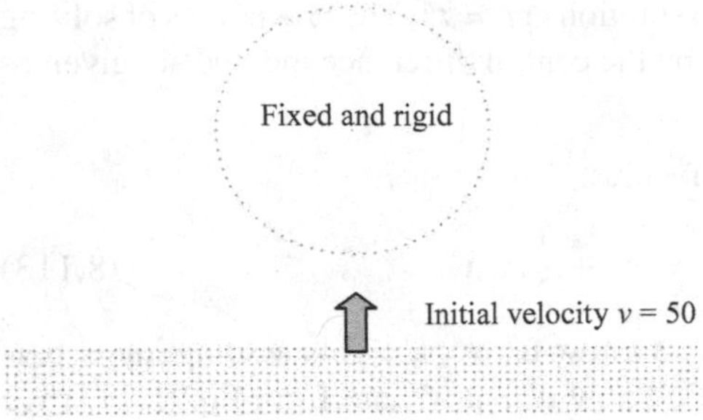

FIGURE 8.1. A beam impacting with a rigid cylinder with an initial velocity (meshless model).

where $\omega_{\max}$ is the maximum frequency of the corresponding linearized dynamic system. The suggested Δt equals $c\Delta t_c$ with $0.8 < c < 0.98$ (Belytschko et al., 2000b).

We now consider the impact between two bodies: one is a rigid cylinder (radius = 0.4 m) and the other is a deformable beam (0.2 m × 2 m), as shown in Figs. 8.1 and 8.2. There are two cases: (1) the cylinder is fixed in space and is impacted by a beam with an initial velocity $v^0 = 50$ m/s and (2) the fixed end beam is impacted by the cylinder moving with a constant velocity $v^c = 100$ m/s. The relevant input data are

$$E = 0.21 \times 10^{12}\ \text{N/m}^2, \quad \upsilon = 0.28, \quad \sigma_Y = 0.2 \times 10^9\ \text{N/m}^2,$$
$$H = 0.21 \times 10^9 \text{N/m}^2, \quad c = 1, \quad \rho^0 = 0.785 \times 10^4\ \text{Kg/m}^3,$$

where E is the Young modulus, υ the Poisson ratio, σ_Y the von Mises strength, H a constant representing the slope of the stress–strain relation in plastic loading, and c a constant with $0 \le c \le 1$, with $c = 1$ implies $\dot{\beta} = 0$, which is the case of isotropic hardening; and $c = 0$ implies that $\bar{\beta}$ does not affect the size of the yield surface—it is the case of kinematic hardening; ρ^0 is the mass density. To test the applicability of the numerical implementation of meshless method on large-strain high-speed impact/contact simulation, very high-speed impact velocity is used in each case. It is noted that these velocities are higher than those in the case of regular car crash. This then demonstrates the applicability of meshless method on the simulation of high-speed impact in general sense.

Case 1. A beam impacting with a rigid cylinder with initial velocity 50 m/s.

This case is concerned with the beam impacting with a rigid cylinder at an initial velocity as illustrated in Fig. 8.1. The objective is to test the contact/separation algorithm during large-strain elastic–plastic deformation. Some of the numerical results of Case 1 are shown in Fig. 8.2a–d. They are the snap shots of von Mises stress distribution on deformed model during the contact/separation process: initial contact, full contact, initial separation, and full separation. As can be seen from the results that the numerical implementation of meshless method on dynamic, large strain, elastic–plastic, impact/contact/separation is successful.

Case 1. The rigid cylinder impacting with a beam with constant velocity 100 m/s.

In this case, the beam is fixed at the two ends and the rigid cylinder is impacting with the beam with a constant high-speed velocity of 100 m/s, as shown in Fig. 8.3.

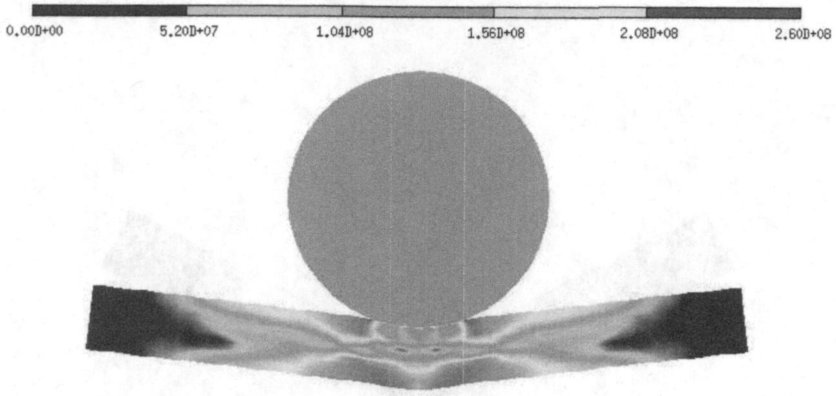

(a) Initial contact (meshless result)

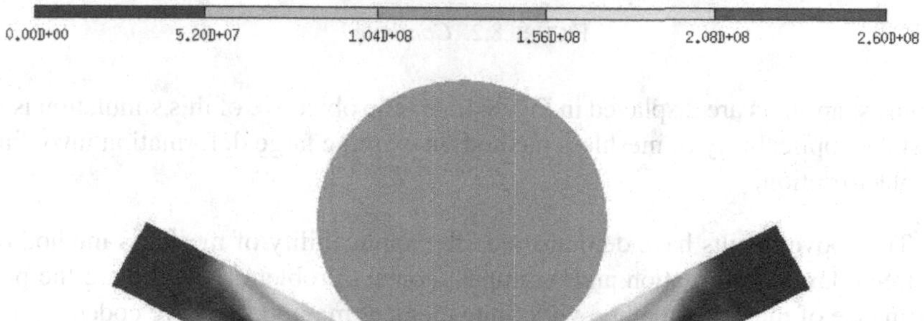

(b) Full contact (meshless result)

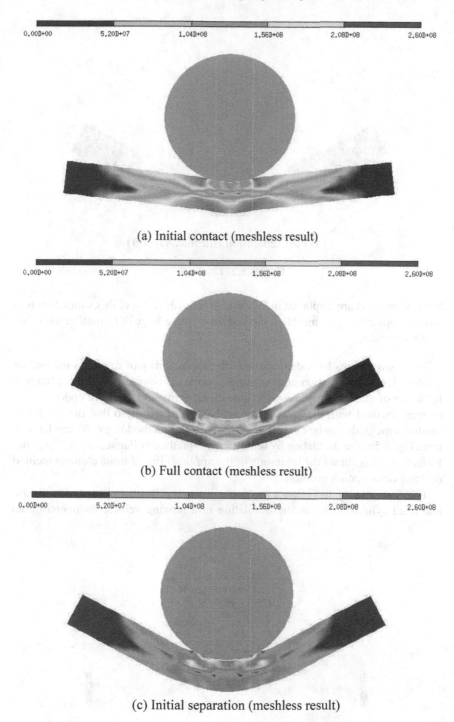

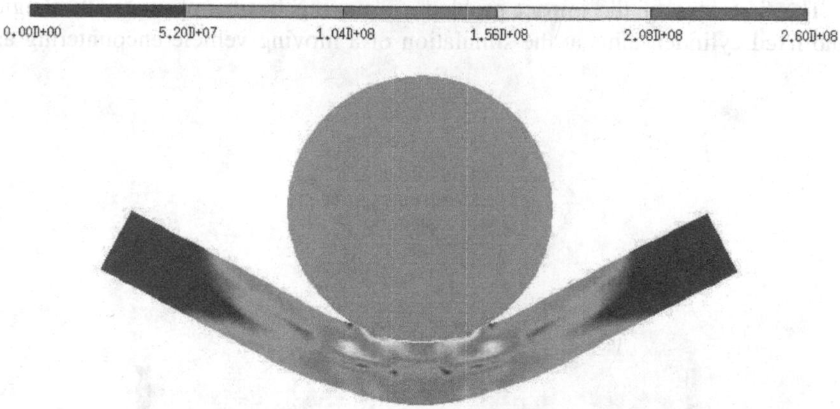

(c) Initial separation (meshless result)

FIGURE 8.2. Von Mises stress distribution during impact/contact/separation (impact with initial velocity).

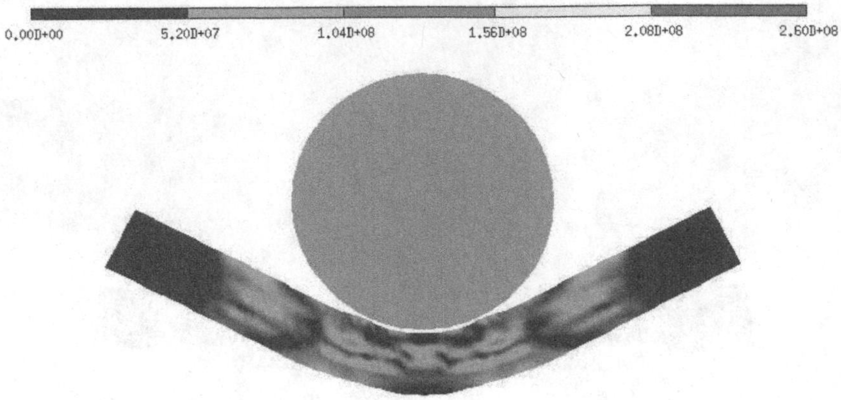

(d) Full separation (meshless result)

FIGURE 8.2. *Continued*

Some snap shots are displayed in Fig. 8.4a–e. The objective of this simulation is to test the applicability of meshless method on extreme large deformation involving contact/friction.

The above results have demonstrated the applicability of meshless method on extreme large deformation and on impact/contact problem. To compare the performance of meshless method with finite element method, we have coded a finite element method with the same exact analytical treatment so that the mesh and meshless methods can be compared in the aspect of methodology. As can be seen from Fig. 8.5b, the simulation by finite element method collapsed at very large deformation. This shows the limitation of the applicability of finite element method on large deformation problem.

The first case of the impact problem, a moving beam impacts with a rigid and fixed cylinder, aims at the simulation of a moving vehicle encountering an

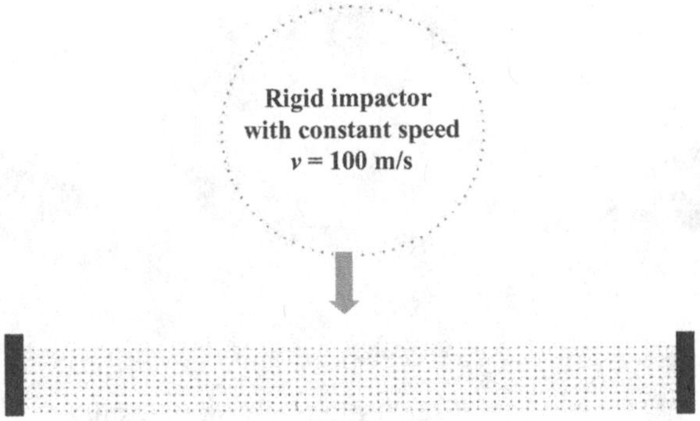

FIGURE 8.3. A rigid cylinder is impacting with a beam with an initial velocity (meshless model).

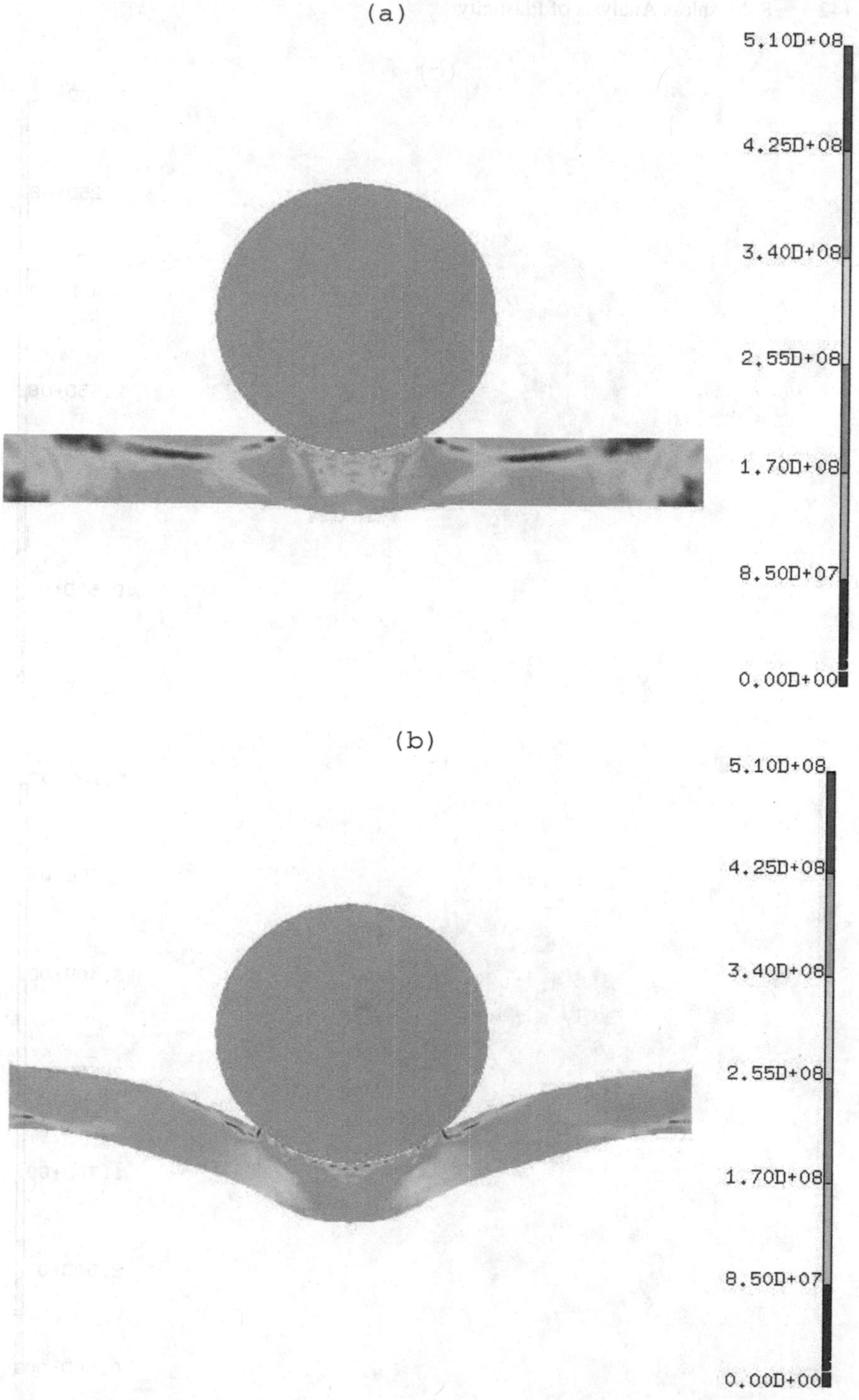

FIGURE 8.4. von Mises stress distribution during (a) initial impact, (b) impact, (c) impact with large deformation, (d) impact with very large deformation, and (e) impact with extreme large deformation.

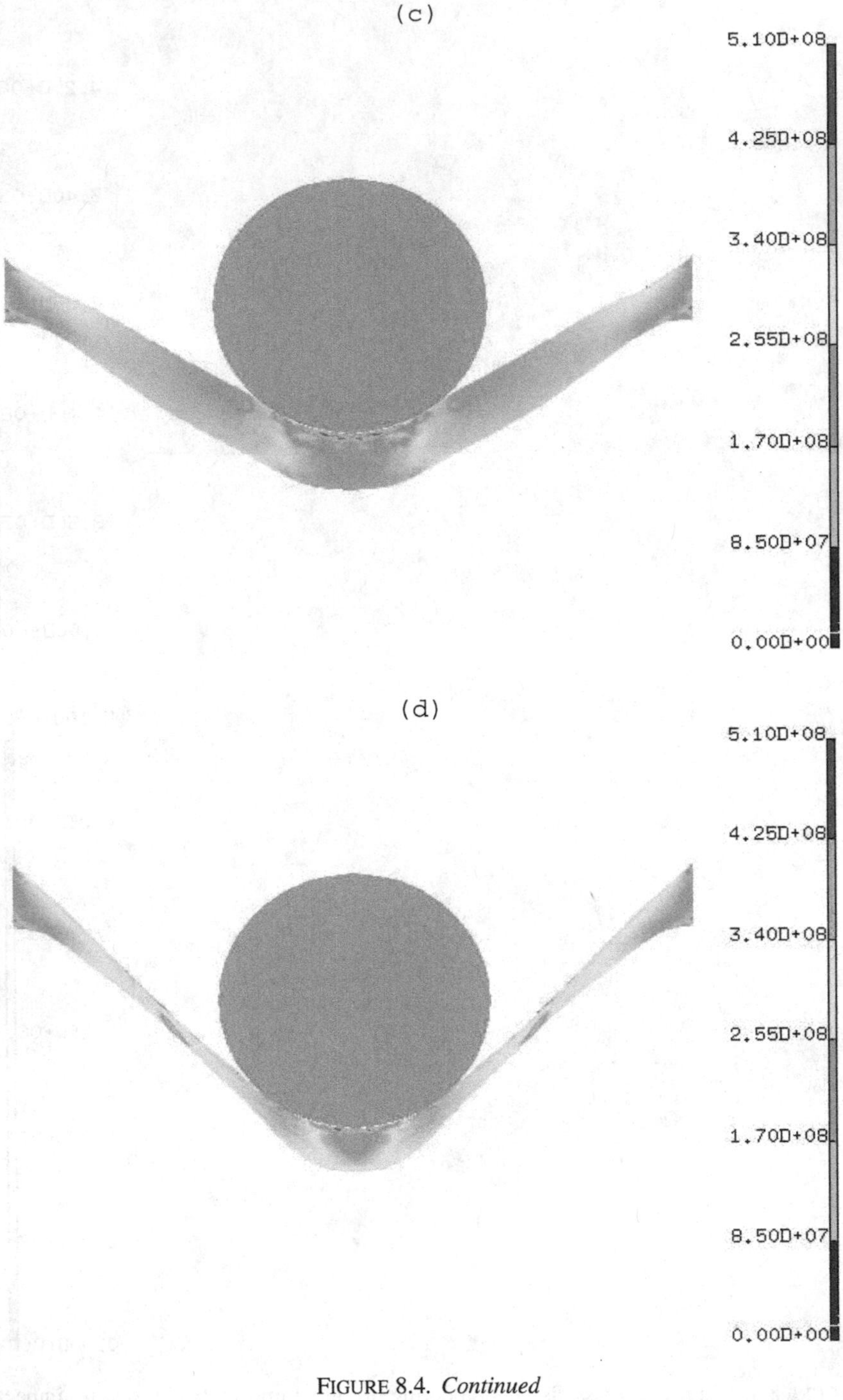

FIGURE 8.4. *Continued*

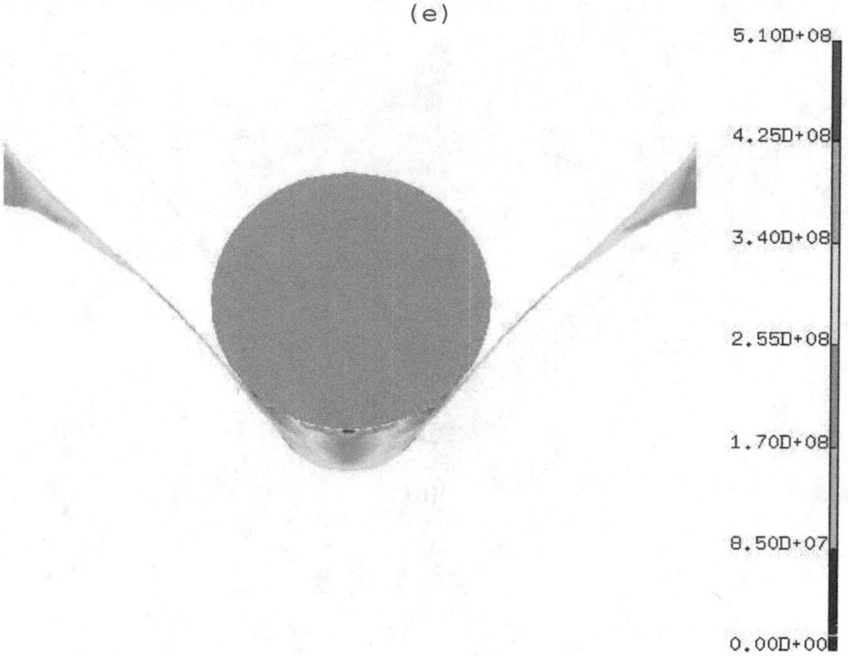

FIGURE 8.4. *Continued*

obstacle, while the second case, a moving rigid cylinder impacts on a fixed end beam, belongs to a general category of metal forming problems. A much more realistic but difficult problem would be the analysis of impact process between two deformable bodies.

Both extreme large deformation and contact/separation modeling have been a research subject pursued by many researchers. The combination is more challenging and is often involved in crash and impact problems. The two numerical cases presented have demonstrated a very promising feature of the application of meshless method on crash and impact simulations.

Incremental Plasticity and Slow Crack Growth Problem

From now on, we focus our attention on quasi-static problems of plasticity with small strain approximations. Recall the governing equation, Eq. (6.35), of the static problems in elasticity from Chapter 6 and recast it for the quasi-static problems as

$$\begin{vmatrix} \mathbf{K} & \mathbf{G} \\ \mathbf{G}' & 0 \end{vmatrix} \begin{vmatrix} \Delta\mathbf{U} \\ \Delta\Lambda \end{vmatrix} = \begin{vmatrix} \Delta\mathbf{F} \\ \Delta\mathbf{f} \end{vmatrix}, \tag{8.122}$$

(a)

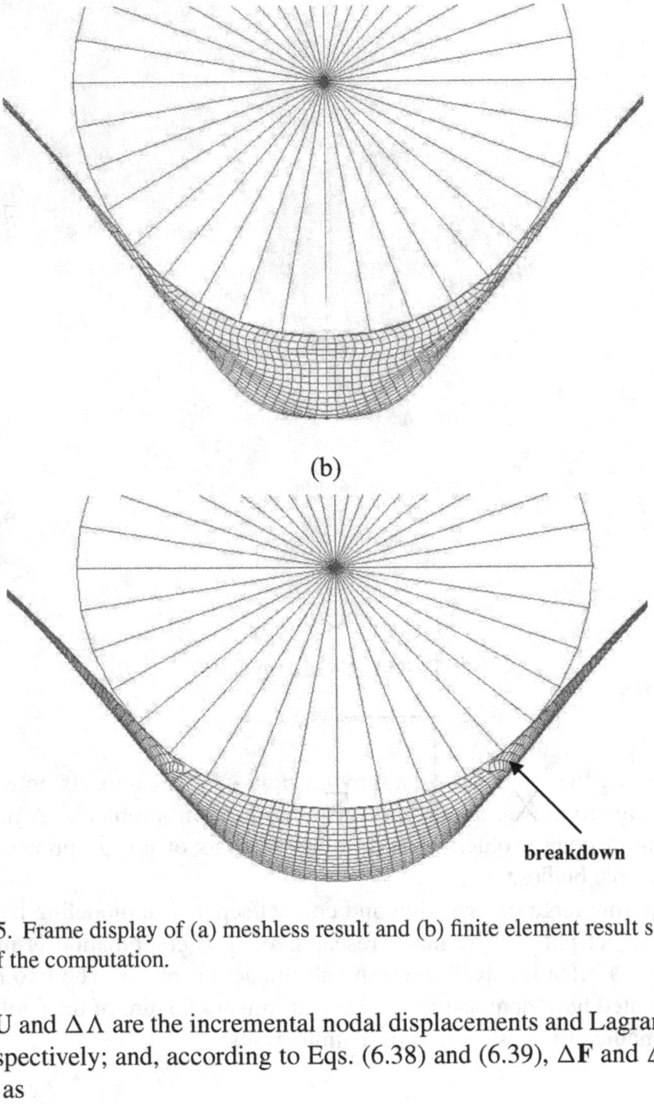

FIGURE 8.5. Frame display of (a) meshless result and (b) finite element result showing the collapse of the computation.

where $\Delta\mathbf{U}$ and $\Delta\mathbf{\Lambda}$ are the incremental nodal displacements and Lagrange multipliers, respectively; and, according to Eqs. (6.38) and (6.39), $\Delta\mathbf{F}$ and $\Delta\mathbf{f}$ may be rewritten as

$$\Delta F_\alpha = \int_{\Gamma_t} \Delta\bar{t}_j \phi_{j\alpha} \mathrm{d}\Gamma, \tag{8.123}$$

$$\Delta f_\alpha = \int_{\Gamma_u} \Delta\bar{u}_j \psi_{j\alpha} \mathrm{d}\Gamma. \tag{8.124}$$

However, in plasticity, $\Delta\mathbf{F}$ should include an additional term due to the application of the return mapping algorithm. It is further discussed as follows. First, we note that, by writing

$$\int_\Omega t_{ij} B_{ij\alpha} \mathrm{d}\boldsymbol{\Omega} = F_\alpha^*, \tag{8.125}$$

$\mathbf{F}^*$ is the nodal force due to the stress tensor. If the stress tensor is further related to strain tensor as

$$t_{ij} = A_{ijkl} e_{kl}, \tag{8.126}$$

then we obtain (cf. Eq. (6.36))

$$F_\alpha^* = \left\{ \int_\Omega A_{ijkl} B_{ij\alpha} B_{kl\beta} \mathrm{d}\Omega \right\} U_\beta = K_{\alpha\beta} U_\beta. \tag{8.127}$$

In other words, in solving problem in plasticity, the first step is to create a trial elastic state and it is equivalent to employing

$$\Delta t_{ij} = A_{ijkl} \Delta \tilde{e}_{kl}, \tag{8.128}$$

and utilizing Eqs. (8.122), (8.103), and (8.124) to obtain the increment displacements and, subsequently, the incremental strains. If the yield function evaluated at the trial elastic state is less than or equal to zero, the process is elastic and the trial state is taken as the actual state. On the other hand, if the trial state violates the Kuhn–Tucker conditions and the return mapping algorithm is used to update the stresses and all the internal variables, it is emphasized that the stresses are not simply calculated by Eq. (8.128); $\mathbf{F}^*$ calculated by Eqs. (8.125) and (8.127) is very much different. The differences should be considered as the residual nodal force, which will be balanced by iteratively solving Eq. (8.122) for $\Delta\mathbf{U}$ and $\Delta\boldsymbol{\Lambda}$ with $\Delta\mathbf{f} = 0$ and

$$\Delta F_\alpha = \int_\Omega t_{ij} B_{ij\alpha} \mathrm{d}\Omega - K_{\alpha\beta} U_\beta. \tag{8.129}$$

Fracture mechanics deals with the rupture of solids in the presence of cracks. Linear elastic fracture mechanics (LEFM) was originally developed to describe crack growth and fracture under essentially elastic conditions, as the name suggests. However, such conditions are met only for plane strain fracture of high-strength metallic materials and for fracture of intrinsically brittle materials like glasses, ceramics, and rocks. Later, it was shown that LEFM concepts could be slightly altered to cope with limited plasticity in the crack tip regime. Nevertheless, there are many important classes of materials that are too ductile to permit description of their behavior by LEFM. It is observed that cracked specimens made of ductile materials subjected to monotonically and slowly increasing load show a considerable amount of crack tip plasticity and nonnegligible amount of stable crack growth prior to the onset of fast fracture.

During the process of crack growth, even in the case of two-dimensional self-similar crack growth, the crack size becomes a monotonically increasing but unknown variable. Precisely speaking, one more unknown variable corresponds to the need of one more governing equation for the system. In order to establish the governing equation for the crack size, various kinds of relations between the crack size and the other fracture parameters have been proposed and tested (Lee et al., 1996, 1997). Here, we take the experiment curve relating the applied load with

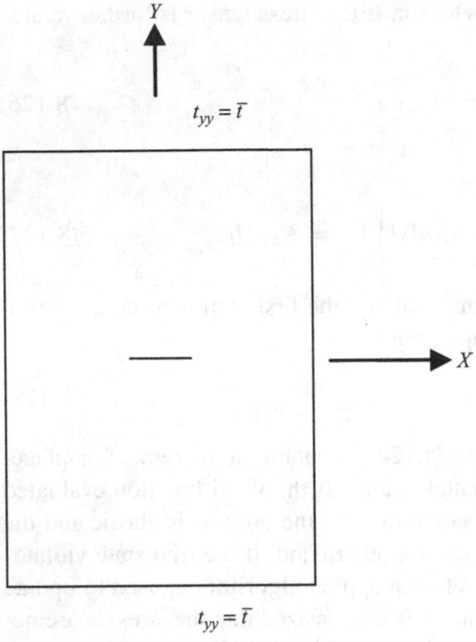

FIGURE 8.6. Center-cracked specimen subjected to mode-I tensile stress.

the incremental crack size as an input and calculate the displacements, strains, stresses, internal variables, and plastic energy as outputs.

A rectangular plate of length $2L$, width $2W$, and thickness B with a centered line crack of initial crack size $2a$ subjected to symmetric boundary conditions is shown in Fig. 8.6. Therefore, only the first quadrant of the plate $R = \{x, y | 0 \leq x \leq W, 0 \leq y \leq L\}$ needs to be analyzed. The specimen is made of 2024-T3 aluminum alloy. A realistic experimental curve relating the applied load $\bar{t}$ and crack size a is shown in Fig. 8.7. The J_2 flow theory with plane stress condition is used to

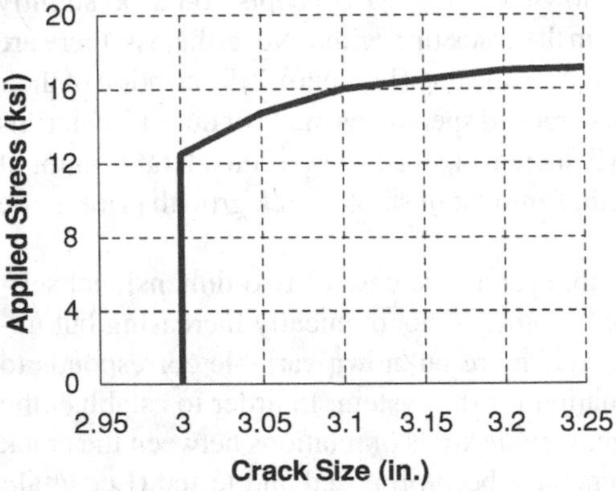

FIGURE 8.7. Applied stress ($t_{yy} = \bar{t}$) vs. crack size (a).

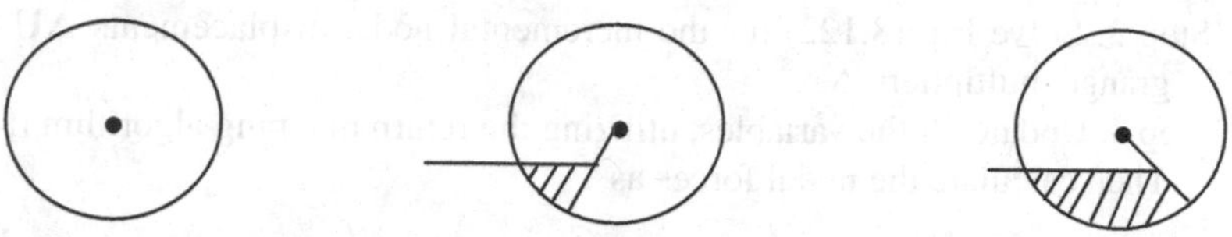

(a) no crack in the domain of influence (b) the presence of a crack (c) crack growth

FIGURE 8.8. Illustration of visibility test.

describe the material behavior of the cracked specimen. The boundary conditions may be specified as

$$t_{yy} = \bar{t}, t_{xy} = 0 \text{ on } y = L, \quad 0 \le x \le W, \tag{8.130}$$

$$t_{xx} = t_{xy} = 0 \text{ on } x = W, \quad 0 \le y \le L, \tag{8.131}$$

$$t_{yy} = t_{xy} = 0 \text{ on } y = 0, \quad 0 < x \le a, \tag{8.132}$$

$$u_y = t_{xy} = 0 \text{ on } y = 0, \quad 0 \le x \le W, \tag{8.133}$$

$$u_x = t_{xy} = 0 \text{ on } x = 0, \quad 0 \le y \le L. \tag{8.134}$$

It is seen that the essential boundary conditions $u_y = 0$ along $y = 0$, $x \in [a, W]$, and $u_x = 0$ along $x = 0$, $y \in [0, L]$ are the consequences of the mirror symmetries that the first quadrant of the specimen has. Also, the line crack $\{x, y | y = 0, 0 \le x \le a\}$ is a barrier. In finite element analysis, the presence of mirror symmetries or barriers does not present a major difficulty. On the contrary, in meshless analysis, the presence of mirror symmetry and/or barrier needs special treatment as indicated in the previous chapter. Moreover, in crack growth problem, the size of the crack, i.e., the geometry of the barrier, is a variable. This will make the problem even more difficult to solve. Figure 8.8 demonstrates that the presence of a crack affects the "visibility" of region near the crack tip and, as crack tip advances, the effect is even more pronounced. The essential step-by-step procedures for solving the slow crack growth problem by using the meshless method are listed as follows.

Step 1. Initialize the stresses, strains, nodal displacements, Lagrange multipliers, and all the internal variables. Set $n = 1$.

Step 2. Based on the current crack size, determine the shape functions $\phi_{j\alpha}$ and $\varphi_{j\alpha}$ and the derivatives of φ, i.e., $B_{ij\alpha} \equiv (\phi_{i\alpha,j} + \phi_{j\alpha,i})/2$. Form matrices **K** and **G** according to Eqs. (6.36) and (6.37), respectively. Calculate the forcing term

$$F_\alpha = \int_\Omega \rho f_i \phi_{i\alpha} \mathrm{d}\Omega + \int_{\Gamma_t} \bar{t}_j(n) \phi_{j\alpha} \mathrm{d}\Gamma, \tag{6.38*}$$

$$f_\alpha = \int_{\Gamma_u} \bar{u}_j(n) \psi_{j\alpha} \mathrm{d}\Gamma, \tag{6.39*}$$

$$\Delta F_\alpha = \int_{\Gamma_t} \{\bar{t}_j(n) - \bar{t}_j(n-1)\} \phi_{j\alpha} \mathrm{d}\Gamma, \tag{6.123*}$$

$$\Delta f_\alpha = \int_{\Gamma_u} \{\bar{u}_j(n) - \bar{u}_j(n-1)\} \psi_{j\alpha} \mathrm{d}\Gamma. \tag{6.124*}$$

Step 3. Solve Eq. (8.122) for the incremental nodal displacements $\Delta\mathbf{U}$ and Lagrange multipliers $\Delta\Lambda$.

Step 4. Update all the variables, utilizing the return mapping algorithm if needed. Then calculate the nodal forces as

$$\hat{F}_\alpha = \int_\Omega t_{ij} B_{ij\alpha} \mathrm{d}\Omega + G_{\alpha\beta}\Lambda_\beta + \int_\Omega \rho f_i \phi_{i\alpha} \mathrm{d}\Omega, \tag{8.135}$$

$$\hat{f}_\alpha = G_{\beta\alpha} U_\beta. \tag{8.136}$$

Define the error as

$$\mathbf{e} \equiv ||\hat{\mathbf{F}} - \mathbf{F}|| + ||\hat{\mathbf{f}} - \mathbf{f}||. \tag{8.137}$$

If $\mathbf{e}$ is less than an error tolerance, then increase n by one and set the crack size at $a(n)$ and go to Step 2; otherwise, set

$$\Delta\mathbf{F} = \hat{\mathbf{F}} - \mathbf{F}, \quad \Delta\mathbf{f} = \hat{\mathbf{f}} - \mathbf{f}, \tag{8.138}$$

and go to Step 3.

A general purpose computer software for meshless analysis of two-dimensional, plane strain and plane stress, plasticity problems has been developed. It is capable of solving crack growth problems. The special treatment for the presence of mirror symmetries and barriers is incorporated. In this work, a center-cracked specimen (cf. Fig. 8.6) subjected to monotonically increasing load (cf. Fig. 8.7) has been analyzed. The relevant input data are:

$$L = 6.0 \text{ in.}, \quad W = 6.0 \text{ in.}, \quad a^0 = 3.0 \text{ in.}, \quad B = 0.062 \text{ in.}, \tag{8.139}$$

$$E = 10,300 \text{ ksi}, \quad \upsilon = 0.33, \quad \sigma_Y = 55 \text{ ksi}, \quad H = 250 \text{ ksi}, \tag{8.140}$$

and, for the case of plane strain, a combined isotropic/kinematic hardening rule, $c = 0.5$, is assumed. The applied loading $t_{yy} = \bar{t}$ increases from 12.5 ksi at $a^0 =$ 3.0 in. to 17.1 ksi at $a = 3.25$ in. From the geometry of the specimen, this is considered as a plane stress problem. As output, for plane stress case, plastic energy, P, defined as

$$P = \int_\Omega \left\{ \int_0^t t_{ij} \dot{\varepsilon}_{ij}^{\mathrm{p}} \mathrm{d}\tau \right\} \mathrm{d}\Omega, \tag{8.141}$$

is shown as a function of the crack size in Fig. 8.9; the stress distributions of t_{yy} are shown in Fig. 8.10. For illustrative purpose, we use the same input data (Fig. 8.7) for plane strain case and the output data are shown in Figs. 8.11 and 8.12.

In fracture mechanics, crack size is an ever-increasing variable, so is plastic energy due to the second law of thermodynamics. Therefore, plastic energy should be a monotonic function of crack size and in this sense Fig. 8.9 has a profound physical meaning. The relation between plastic energy and crack size being linear makes it possible to serve as a criterion for slow crack growth process (Lee et al., 1996, 1997). Figure 8.10a,b clearly indicates that, as crack grows, the plastic region and the stresses in front of the crack tip increase and in the wake

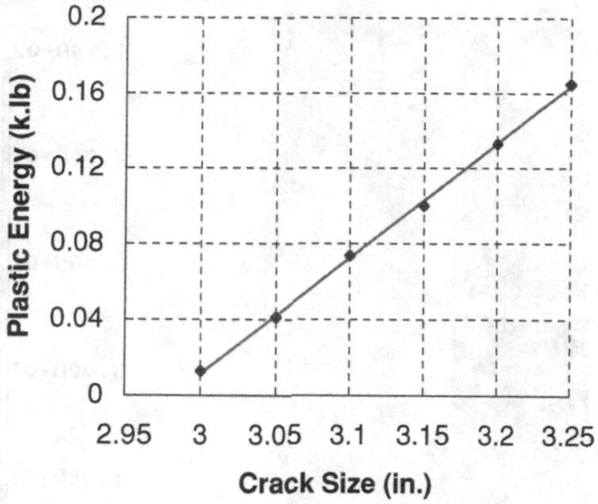

FIGURE 8.9. Plastic energy vs. crack size.

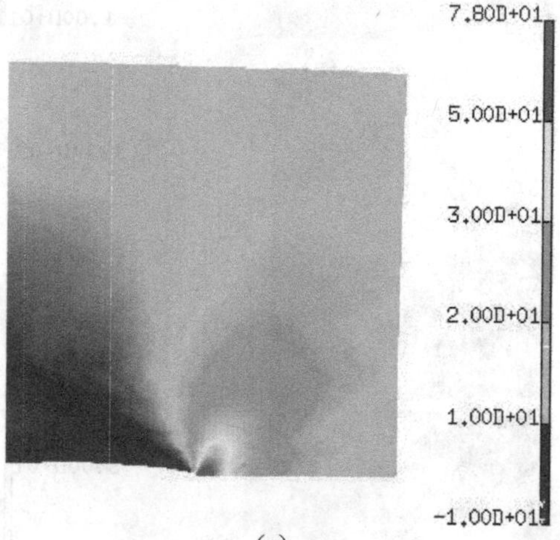

(a)

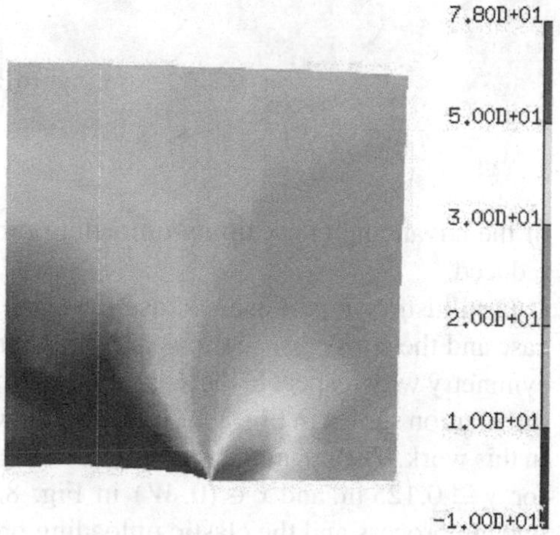

(b)

FIGURE 8.10. (a) Stress distribution (t_{yy}) at initial crack size $a = 3.0$ in. (plane stress case) and (b) Stress distribution (t_{yy}) as crack size grows to be $a = 3.25$ in. (plane stress case).

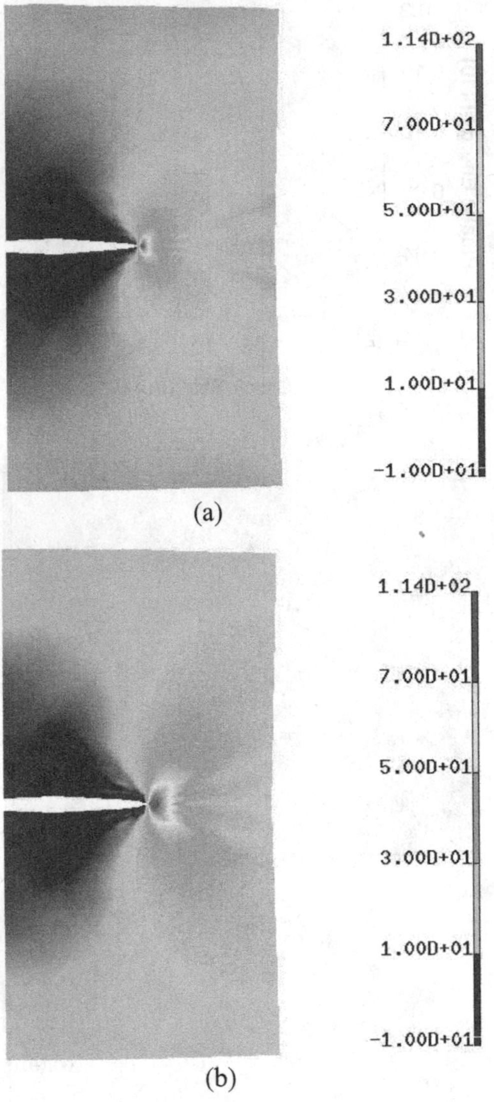

FIGURE 8.11. (a) Stress distribution (t_{yy}) at initial crack size $a = 3.0$ in. (plane strain case), (b) stress distribution (t_{yy}) as crack size grows to be $a = 3.25$ in. (plane strain case).

of the advancing crack tip an unloading process is ongoing and the stresses are reduced.

For illustrative purpose, we use the same input data (Fig. 8.7) for plane strain case and the stress distributions are shown in Fig. 8.11a,b. In this case, the mirror symmetry with respect to the x–z plane was not utilized, the symmetry of the stress distribution shown in Fig. 8.11 indicates the validity of the computer software used in this work. The opening stresses t_{yy} for $a = 3.00$ in. and $a = 3.25$ in. are plotted for $y \cong 0.125$ in. and $x \in (0, W)$ in Fig. 8.12, which also indicates the plastic loading process and the elastic unloading process near the advancing crack tip.

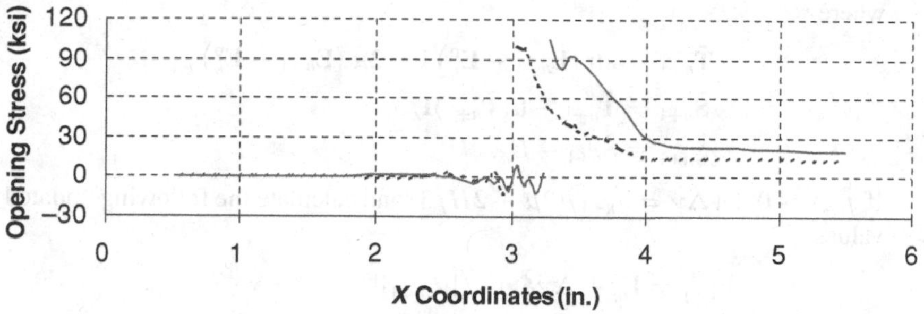

FIGURE 8.12. Stress distribution (t_{yy}) along the line of crack (plane strain case). Solid line: $a = 3.25$ in., dot line: $a = 3.0$ in.

Remarks

In this chapter, we formulate a general theory of plasticity in an axiomatic approach with full utilization of the concept of internal variables and the constraint due to Clausius–Duhem inequality. The two example problems solved demonstrated that this theory is valid for both static and dynamic cases with either finite or infinitesimal strains. The return mapping algorithm works exactly and analytically if the J_2 flow theory is adopted. It is also shown that the problems with mirror symmetry and/or moving barrier can be rigorously and successfully treated.

For the readers who want to gain knowledge through hands-on experience, a meshless computer program for the analysis of slow crack growth in elastoplastic continuum can be found on the book's page at http://www.springeronline.com/.The corresponding user's manual is included in Appendix E

Problems

1. The J_2 model for three-dimensional and plane strain problems is defined by Eqs. (8.67)–(8.70). (a) Show that this model is an associative model of plasticity. (b) Prove that $\Delta\gamma$ given by Eq. (8.74) is the solution of the initial value problem specified by Eqs. (8.57)–(8.60).
2. The J_2 model for plane stress problem is defined by Eqs. (8.84)–(8.87). (a) Show that this model is an associative model of plasticity. (b) Prove that $\Delta\gamma \equiv \bar{\gamma}/E$ obtained by solving the quartic Eq. (8.95) satisfies Eq. (8.94) and is the solution of the initial value problem specified by Eqs. (8.57)–(8-60).
3. Let the Lagrangian strain at t_{n+1} be $\mathbf{E}_{n+1}$ and let the trial value of the yield function be

$$\tilde{f}_{n+1} = ||\tilde{\boldsymbol{\xi}}_{n+1}|| - \sqrt{2/3}(\sigma_Y + cH\bar{\beta}_n),$$

where

$$\begin{aligned}
\tilde{\mathbf{T}}_{n+1} &= \lambda \mathrm{tr}\left(\mathbf{E}_{n+1} - \mathbf{E}_n^{\mathrm{p}}\right)\mathbf{I} + 2\mu\left(\mathbf{E}_{n+1} - \mathbf{E}_n^{\mathrm{p}}\right), \\
\tilde{\mathbf{S}}_{n+1} &= \tilde{\mathbf{T}}_{n+1} - \mathrm{tr}(\tilde{\mathbf{T}}_{n+1})\mathbf{I}/3, \\
\tilde{\boldsymbol{\xi}}_{n+1} &\equiv \tilde{\mathbf{S}}_{n+1} - \boldsymbol{\beta}_n.
\end{aligned}$$

If $\tilde{f}_{n+1} > 0$, let $\Delta\gamma = \tilde{f}_{n+1}/(2\mu + 2H/3)$ and calculate the following updated values

$$\begin{aligned}
\mathbf{E}_{n+1}^{\mathrm{p}} &= \mathbf{E}_n^{\mathrm{p}} + \Delta\gamma\tilde{\boldsymbol{\xi}}_{n+1}/||\tilde{\boldsymbol{\xi}}_{n+1}||, \\
\boldsymbol{\beta}_{n+1}^{\mathrm{p}} &= \boldsymbol{\beta}_n^{\mathrm{p}} + 2(1-c)H\Delta\gamma\tilde{\boldsymbol{\xi}}_{n+1}/||\tilde{\boldsymbol{\xi}}_{n+1}||/3, \\
\bar{\beta}_{n+1} &= \bar{\beta}_n + \sqrt{2/3}\Delta\gamma, \\
\mathbf{T}_{n+1} &= \lambda \mathrm{tr}\left(\mathbf{E}_{n+1} - \mathbf{E}_{n+1}^{\mathrm{p}}\right)\mathbf{I} + 2\mu\left(\mathbf{E}_{n+1} - \mathbf{E}_{n+1}^{\mathrm{p}}\right), \\
\mathbf{S}_{n+1} &= \mathbf{T}_{n+1} - \mathrm{tr}(\mathbf{T}_{n+1})\mathbf{I}/3.
\end{aligned}$$

Prove that, with these updated values, the value of the yield function indeed returns to zero, i.e.,

$$f_{n+1} = ||\mathbf{S}_{n+1} - \boldsymbol{\beta}_{n+1}|| - \sqrt{2/3}(\sigma_Y + cH\bar{\beta}_{n+1}) = 0.$$

4. The simple tension-compression problem is defined as the one with deformation gradient being constant in space and all stress components vanishing except $t_{11} = \sigma$. Based on the constitutive relations of plasticity given in this chapter, plot the trajectory of stress and strain in case of simple tension-compression including loadings and unloadings.

Appendix A
Vectors and Tensors

Scalars, Vectors, and Tensors

There are many physical quantities with which only a single magnitude can be associated. For example, the mass density may vary throughout the bulk of a material, but in the neighborhood of a given point, it is found to be a constant. We may associate this density with the point. There is no sense of direction associated with the density. Such quantities are called *scalars* and in any system of units they are specified by a single real number.

There are other quantities associated with a point that have not only a magnitude but also a direction. If a force of 1 kg is said to act at a certain point, it is not fully specified until the direction is given. Such a physical quantity is a *vector*. Vectors are entities possessing both magnitude and direction and obeying certain laws. Examples include force, velocity, and acceleration. A vector can be expressed with respect to a frame of reference with three base vectors as

$$\boldsymbol{v} = a\boldsymbol{i} + b\boldsymbol{j} + c\boldsymbol{k}, \tag{A.1}$$

where $\boldsymbol{i}, \boldsymbol{j}$, and $\boldsymbol{k}$ are the base vectors and a, b, and c the components. In ordinary three-dimensional space, the system defined by three mutually orthogonal directions with equal units of measurements is called *Cartesian*. The base vectors may be thought of as lines of unit length lying along the three axes.

The word *tensor* is quite general and when necessary its order must be specifically mentioned, for it will appear that a scalar is a tensor of order zero and a vector is a tensor of order 1. A 3×3 matrix can be written down in a tensor form as a tensor of order 2. Physical quantities are rarely associated with tensors of higher order than the second, but tensors up to the fourth order will arise.

Example 1. The Kronecker delta δ_{ij} is a second-order tensor and is frequently used in the continuum mechanics:

$$\delta_{ij} = \begin{cases} 1, & i = j, \\ 0, & i \neq j. \end{cases} \tag{A.2}$$

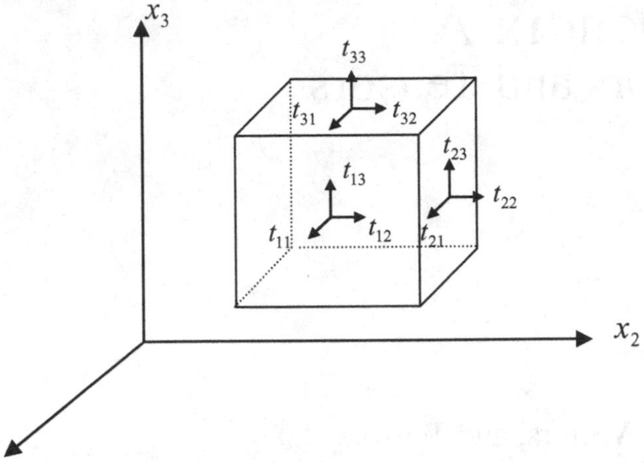

FIGURE A.1. Stress tensor

Example 2. Stress tensor t_{kl} is another most common tensor in continuum mechanics, as shown in Fig. A.1, where k indicates the coordinate surface $x_k =$ constant and l the direction.

Vector Calculus

Any two vector quantities of the same kind (e.g., two forces or two velocities) may be represented as two vectors $\boldsymbol{a}$ and $\boldsymbol{b}$ placed in such a way that the initial point of $\boldsymbol{b}$ coincides with the terminal point of $\boldsymbol{a}$, as in Fig. A.2. The sum of $\boldsymbol{a}$ and $\boldsymbol{b}$ is then defined as the vector $\boldsymbol{c}$ extending from the initial point of $\boldsymbol{a}$ to the terminal point of $\boldsymbol{b}$, $\mathbf{c} = \mathbf{a} + \mathbf{b}$, as shown in Fig. A.2.

The actual addition of two vectors is most conveniently performed by using their rectangular components in some coordinate system, but the definition and properties of vectors addition do not depend on the introduction of a coordinate system.

From the definition, it follows that

$$\boldsymbol{b} + \boldsymbol{a} = \boldsymbol{a} + \boldsymbol{b}, \tag{A.3}$$

as is indicated by the dashed arrows in Fig. A.2. Thus, vector addition obeys the commutative law. It also follows from the definition that the associative law

$$(\boldsymbol{a} + \boldsymbol{b}) + \boldsymbol{c} = \boldsymbol{a} + (\boldsymbol{b} + \boldsymbol{c}) \tag{A.4}$$

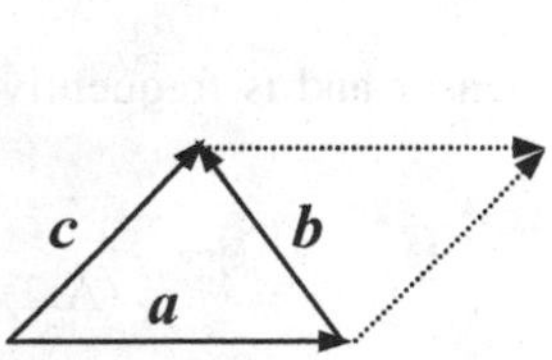

FIGURE A.2. Addition of Vectors

is satisfied. The common value of the two expressions is written as the sum of the three vectors, $\boldsymbol{a}+\boldsymbol{b}+\boldsymbol{c}$.

If a vector is reversed in direction with no change in magnitude, the resulting vector is called the negative of the original vector. To subtract $\boldsymbol{b}$ from $\boldsymbol{a}$, add the negative of $\boldsymbol{b}$ to $\boldsymbol{a}$:

$$\boldsymbol{a}-\boldsymbol{b}=\boldsymbol{a}+(-\boldsymbol{b}). \tag{A.5}$$

A vector $\boldsymbol{a}$ may be multiplied by a scalar c to yield a new vector, $c\boldsymbol{a}$ or $\boldsymbol{a}c$, of magnitude of $|ca|$. If c is positive, $c\boldsymbol{a}$ has the same direction as $\boldsymbol{a}$; if c is negative, $c\boldsymbol{a}$ has the direction of $-\boldsymbol{a}$. If c is zero, the product is a zero vector with zero magnitude and undefined direction.

There are two different kinds of multiplication of one vector by another. In one case, the product is not a vector but a scalar; it is called the *scalar product* or *dot product*. The second case yields a vector; it is called the *vector product* or *cross product*.

The scalar product is defined as the product of two magnitudes and the cosine of the angle between the vectors:

$$\boldsymbol{a}\cdot\boldsymbol{b}=ab\cos\theta. \tag{A.6}$$

It follows immediately from the definition that if m and n are scalars, then

$$(m\boldsymbol{a})\cdot(n\boldsymbol{b})=mn(\boldsymbol{a}\cdot\boldsymbol{b}). \tag{A.7}$$

Apparently, the scalar product is commutative.

$$\boldsymbol{a}\cdot\boldsymbol{b}=\boldsymbol{b}\cdot\boldsymbol{a}. \tag{A.8}$$

The scalar product is also distributive, i.e.,

$$\boldsymbol{a}\cdot(\boldsymbol{b}+\boldsymbol{c})=(\boldsymbol{a}\cdot\boldsymbol{b})+(\boldsymbol{a}\cdot\boldsymbol{c}). \tag{A.9}$$

if

$$\boldsymbol{a}=a_x\boldsymbol{i}+a_y\boldsymbol{j}+a_z\boldsymbol{k},\quad \boldsymbol{b}=b_x\boldsymbol{i}+b_y\boldsymbol{j}+b_z\boldsymbol{k},$$

$$\boldsymbol{a}\cdot\boldsymbol{b}=a_xb_x+a_yb_y+a_zb_z, \tag{A.10}$$

or

$$\boldsymbol{a}\cdot\boldsymbol{b}=a_ib_i \tag{A.11}$$

in indicial notation. This is very important formula. It furnishes the method by which we actually evaluate scalar products.

The product of one vector multiplied by another is the vector product or cross product:

$$\begin{aligned}\boldsymbol{c}&=\boldsymbol{a}\times\boldsymbol{b},\\&=(a_yb_z-a_zb_y)\boldsymbol{i}+(a_zb_x-a_xb_z)\boldsymbol{j}+(a_xb_y-a_yb_x)\boldsymbol{k},\end{aligned} \tag{A.12}$$

which may be written as the formal expansion of the determinant

$$\boldsymbol{a} \times \boldsymbol{b} = \begin{vmatrix} i & j & k \\ a_x & a_y & a_z \\ b_x & b_y & b_z \end{vmatrix}. \tag{A.13}$$

Magnitude c is given by

$$c = ab \sin\theta. \tag{A.14}$$

The commutative law does not hold for vector product. We have instead

$$\boldsymbol{b} \times \boldsymbol{a} = -(\boldsymbol{a} \times \boldsymbol{b}). \tag{A.15}$$

The vector product is still distributive

$$\boldsymbol{a} \times (\boldsymbol{b} + \boldsymbol{c}) = (\boldsymbol{a} \times \boldsymbol{b}) + (\boldsymbol{a} \times \boldsymbol{c}). \tag{A.16}$$

However, the vector product is not associative

$$\boldsymbol{a} \times (\boldsymbol{b} \times \boldsymbol{c}) \neq (\boldsymbol{a} \times \boldsymbol{b}) \times \boldsymbol{c}. \tag{A.17}$$

We have

$$\boldsymbol{a} \times (\boldsymbol{b} \times \boldsymbol{c}) = (\boldsymbol{a} \cdot \boldsymbol{c})\boldsymbol{b} - (\boldsymbol{a} \cdot \boldsymbol{b})\boldsymbol{c}. \tag{A.18}$$

$$(\boldsymbol{a} \times \boldsymbol{b}) \times \boldsymbol{c} = (\boldsymbol{a} \cdot \boldsymbol{c})\boldsymbol{b} - (\boldsymbol{b} \cdot \boldsymbol{c})\boldsymbol{a}. \tag{A.19}$$

For a vector function, $\boldsymbol{u}(t)$, the derivative of a vector can be expressed in terms of derivatives of its components as

$$\frac{\mathrm{d}\boldsymbol{u}}{\mathrm{d}t} = \frac{\mathrm{d}u_k}{\mathrm{d}t}\boldsymbol{i}_k. \tag{A.20}$$

It is then easy to see that we can immediately apply our knowledge of differential calculus to a vector function $\boldsymbol{u}(t)$.

$$\frac{\mathrm{d}}{\mathrm{d}t}(\lambda \boldsymbol{u}) = \lambda \frac{\mathrm{d}\boldsymbol{u}}{\mathrm{d}t} + \frac{\mathrm{d}\lambda}{\mathrm{d}t}\boldsymbol{u}, \tag{A.21}$$

$$\frac{\mathrm{d}}{\mathrm{d}t}(\boldsymbol{u} + \boldsymbol{v}) = \frac{\mathrm{d}\boldsymbol{u}}{\mathrm{d}t} + \frac{\mathrm{d}\boldsymbol{v}}{\mathrm{d}t}, \tag{A.22}$$

$$\frac{\mathrm{d}}{\mathrm{d}t}(\boldsymbol{u} \cdot \boldsymbol{v}) = \frac{\mathrm{d}\boldsymbol{u}}{\mathrm{d}t} \cdot \boldsymbol{v} + \boldsymbol{u} \cdot \frac{\mathrm{d}\boldsymbol{v}}{\mathrm{d}t}, \tag{A.23}$$

$$\frac{\mathrm{d}}{\mathrm{d}t}(\boldsymbol{u} \times \boldsymbol{v}) = \frac{\mathrm{d}\boldsymbol{u}}{\mathrm{d}t} \times \boldsymbol{v} + \boldsymbol{u} \times \frac{\mathrm{d}\boldsymbol{v}}{\mathrm{d}t}, \tag{A.24}$$

where λ is a scalar function.

The most important integral theorem of vector analysis is the divergence theorem, which is also called Green–Gauss theorem

$$\int_v \operatorname{div} \boldsymbol{u} \, \mathrm{d}v = \int_v \nabla \cdot \boldsymbol{u} \, \mathrm{d}v = \oint_\Gamma \boldsymbol{u} \cdot \boldsymbol{n} \, \mathrm{d}a. \tag{A.25}$$

The use of divergence theorem is fundamental in the derivation of differential equations in continuum mechanics, where surface integrals involving surface tractions are converted to volume integrals.

The differential operators gradient of a scalar ϕ and the divergence and curl of a vector $\boldsymbol{A}$ are defined as

$$\nabla \equiv \frac{\partial}{\partial x_i} \boldsymbol{i}_i, \tag{A.26}$$

$$\nabla \phi = \phi_{,i} \boldsymbol{i}_i, \tag{A.27}$$

$$\operatorname{div} \boldsymbol{A} \equiv \nabla \cdot \boldsymbol{A} = A_{i,i}, \tag{A.28}$$

$$\operatorname{curl} \boldsymbol{A} = \nabla \times \boldsymbol{A} = e_{ijk} A_{k,j} \boldsymbol{i}_i, \tag{A.29}$$

where $\boldsymbol{i}_k$ is the base vector and e_{ijk} the usual *permutation tensor* ($e_{123} = e_{312} = e_{231} = -e_{213} = -e_{321} = -e_{132} = 1$, all other $e_{ijk} = 0$).

Tensor Analysis

Tensors are a generalization of vector concepts. They provide powerful tools for the formulation of physical laws in a systematic fashion. The tensor $\boldsymbol{A}$, of order n, is a quantity defined by 3^n components, which may be written as $\underbrace{A_{ijklm\ldots}}_{n\ \text{indices}}$, provided that under rotation to a new coordinate frame they transform according to the law

$$A^*_{\alpha\beta\gamma\delta\ldots} = Q_{\alpha a} Q_{\beta b} Q_{\gamma c} Q_{\delta d} \ldots A_{abcd\ldots} (\det \boldsymbol{Q})^N, \tag{A.30}$$

where $\boldsymbol{Q}$ is the matrix of direction cosine. In the case $N = 0$, $\boldsymbol{A}$ is an absolute tensor, and in the case $N = 1, \boldsymbol{A}$ is an axial tensor.

If interchange of two of the indices does not change the value of the component, the tensor is said to be *symmetric* with respect to these indices. If the absolute value is unchanged but the sign is reversed, it is *antisymmetric* with respect to the indices.

Tensor addition and multiplication by a scalar obey the following four addition axioms and four scalar-multiple axioms, characteristic of a generalized vector space.

Addition Axioms:

(a) $\boldsymbol{T} + \boldsymbol{U} = \boldsymbol{U} + \boldsymbol{T}$ (commutative)
(b) $\boldsymbol{T} + (\boldsymbol{U} + \boldsymbol{V}) = (\boldsymbol{T} + \boldsymbol{U}) + \boldsymbol{V}$ (associative)
(c) $\boldsymbol{T} + 0 = \boldsymbol{T}$ (A.31)
(d) $\boldsymbol{T} + (-\boldsymbol{T}) = 0$

Scalar-Multiple Axioms:

(a) $a(b\boldsymbol{T}) = (ab)\boldsymbol{T}$
(b) $1\boldsymbol{T} = \boldsymbol{T}$
(c) $(a + b)\boldsymbol{T} = a\boldsymbol{T} + b\boldsymbol{T}$ (A.32)
(d) $a(\boldsymbol{T} + \boldsymbol{U}) = a\boldsymbol{T} + a\boldsymbol{U}$

The scalar product of two tensors is a scalar, denoted $\boldsymbol{T}:\boldsymbol{U}$, which can be calculated in terms of components of the two tensors in any one rectangular Cartesian system, by

$$\boldsymbol{T}:\boldsymbol{U} = T_{ij}U_{ij}. \tag{A.33}$$

The tensor product or open product of two vectors, denoted $\boldsymbol{ab}$, is a tensor called a *dyad*, defined by the requirement that

$$(\boldsymbol{ab})\cdot\boldsymbol{v} = \boldsymbol{a}(\boldsymbol{b}\cdot\boldsymbol{v}), \tag{A.34}$$

for all vectors $\boldsymbol{v}$. That is, if $\boldsymbol{T} = \boldsymbol{ab}$, then $\boldsymbol{T}\cdot\boldsymbol{v} = \boldsymbol{a}(\boldsymbol{b}\cdot\boldsymbol{v})$ for all vectors $\boldsymbol{v}$. In rectangular Cartesian components,

$$T_{ij} = a_i b_j. \tag{A.35}$$

The product of two second-order tensors, denoted $\boldsymbol{T}\cdot\boldsymbol{U}$, means the composition of the two operations $\boldsymbol{T}$ and $\boldsymbol{U}$, with $\boldsymbol{U}$ performed first, defined by the requirement that

$$(\boldsymbol{T}\cdot\boldsymbol{U})\cdot\boldsymbol{v} = \boldsymbol{T}\cdot(\boldsymbol{U}\cdot\boldsymbol{v}) \tag{A.36}$$

for all vectors $\boldsymbol{v}$. If $\boldsymbol{P} = \boldsymbol{T}\cdot\boldsymbol{U}$, then

$$P_{ij} = T_{ik}U_{kj} \tag{A.37}$$

or $\boldsymbol{P} = \boldsymbol{T}\boldsymbol{U}$ in matrix notation, where $\boldsymbol{P}$, $\boldsymbol{T}$, and $\boldsymbol{U}$ are the matrices of components in any one rectangular Cartesian system.

Tensor Algebra Axioms

$$\begin{aligned}
&\text{(a)}\ (\boldsymbol{T}\cdot\boldsymbol{U})\cdot\boldsymbol{R} = \boldsymbol{T}\cdot(\boldsymbol{U}\cdot\boldsymbol{R})\\
&\text{(b)}\ \boldsymbol{T}\cdot(\boldsymbol{R}+\boldsymbol{U}) = \boldsymbol{T}\cdot\boldsymbol{R}+\boldsymbol{T}\cdot\boldsymbol{U}\\
&\text{(c)}\ (\boldsymbol{R}+\boldsymbol{U})\cdot\boldsymbol{T} = \boldsymbol{R}\cdot\boldsymbol{T}+\boldsymbol{U}\cdot\boldsymbol{T}\\
&\text{(d)}\ a(\boldsymbol{T}\cdot\boldsymbol{U}) = (a\boldsymbol{T})\cdot\boldsymbol{U} = \boldsymbol{T}\cdot(a\boldsymbol{U})\\
&\text{(e)}\ \boldsymbol{I}\cdot\boldsymbol{T} = \boldsymbol{T}\cdot\boldsymbol{I} = \boldsymbol{T}
\end{aligned} \tag{A.38}$$

The differential calculus to a tensor function follows the same way as to a vector function in Eqs. (A.21)–(A.24).

Example 3. Deformation gradients and some identities.
In the description of the deformation and motion of a continuous body, two sets of coordinates may be used: the material or Lagrangian coordinate for undeformed body and the spatial or Eulerian coordinates for deformed body. If we call X_K the Lagrangian coordinate and x_k the Eulerian coordinate, the deformation gradient is defined as

$$x_{k,K} \equiv \frac{\partial x_k}{\partial X_K}, \quad X_{K,k} \equiv \frac{\partial X_K}{\partial x_k}. \tag{A.39}$$

Through the chain rule of partial differential, it is clear that

$$x_{k,K}X_{K,l} = \delta_{kl}, \quad X_{K,k}x_{k,L} = \delta_{KL}. \tag{A.40}$$

Each one of the two sets is a set of nine linear equations for the nine unknowns $x_{k,K}$ or $X_{K,k}$. A unique solution exists, if the Jacobian of the transformation is assumed that it does not vanish. Using Cramer's rule of determinants, the solution for $X_{K,k}$ may be obtained in terms of $x_{k,K}$. Thus,

$$X_{K,k} = \frac{\text{cofactor } x_{k,K}}{j} = \frac{1}{2j} e_{KLM} e_{klm} x_{l,L} x_{m,M}, \tag{A.41}$$

where e_{KLM} and e_{klm} are permutation tensors and

$$j \equiv \left| x_{k,K} \right| = \tfrac{1}{6} e_{KLM} e_{klm} x_{k,K} x_{l,L} x_{m,M} \tag{A.42}$$

is the Jacobian. By differentiating Eq. (A.42), the following identities can be obtained

$$(jX_{K,k})_{,K} = 0, \tag{A.43}$$

$$(j^{-1} x_{k,K})_{,k} = 0, \tag{A.44}$$

$$\frac{\partial j}{\partial x_{k,K}} = \text{cofactor } x_{k,K} = jX_{K,k}. \tag{A.45}$$

The proof of Eq. (A.43) is shown as follows.

$$\begin{aligned}(jX_{K,k})_K &= (\tfrac{1}{6} e_{KLM} e_{klm} x_{l,L} x_{m,M} \delta_{kk})_K, \\ &= \tfrac{1}{2} e_{KLM} e_{klm} (x_{l,L} x_{m,M})_K, \\ &= \tfrac{1}{2} e_{KLM} e_{klm} x_{l,LK} x_{m,M} + \tfrac{1}{2} e_{KLM} e_{klm} x_{l,L} x_{m,MK}.\end{aligned} \tag{A.46}$$

Since

$$e_{KLM} = -e_{LKM}, \quad e_{KLM} = -e_{KML}, \tag{A.47}$$

while

$$x_{l,LK} = x_{l,KL}, \quad x_{m,MK} = x_{m,KM}. \tag{A.48}$$

Hence

$$e_{KLM} x_{l,LK} = e_{KLM} x_{m,MK} = 0. \tag{A.49}$$

Equation (A.43) is then resulted and proved.

Invariants of Vectors and Tensors

The theory of invariants of vectors and tensors is very useful in constructing constitutive equations for scalar, vector, and tensor functions of vector and tensor variables. We shall be concerned only with constitutive equations, which are invariant under the full group of orthogonal transformations of the rectangular frame of reference.

If $\boldsymbol{x}$ is a set of rectangular coordinates, then the transformation

$$x_i^* = Q_{ij} x_j \tag{A.50}$$

determines a new set of rectangular coordinates $\boldsymbol{x}^*$ if Q_{ij} has the property

$$\boldsymbol{Q}\boldsymbol{Q}^T = \boldsymbol{Q}^T\boldsymbol{Q} = 1, \quad \det \boldsymbol{Q} = \pm 1. \tag{A.51}$$

The matrix $\boldsymbol{Q}$ satisfying Eq. (A.51) is said to be an orthogonal transformation. The set of all orthogonal transformations forms a group. This is the full group of orthogonal transformations. The set of orthogonal transformations with positive determinations also forms a group. This is the proper group.

Under transformations of coordinates (A.50), a vector v_i and a second-order tensor A_{ij} transform according to:

$$v_i^* = Q_{ik} v_k, \tag{A.52}$$

$$A_{ij}^* = Q_{ik} Q_{jl} A_{kl}. \tag{A.53}$$

Any axial vector, w_i, which transforms according to

$$w_i^* = Q_{ik} w_k \det \boldsymbol{Q}, \tag{A.54}$$

may be replaced by skew-symmetric tensor W_{ij}

$$W_{ij} = e_{ijk} w_k, \tag{A.55}$$

which transforms like Eq. (A.53).

A function $f(\boldsymbol{v}_1, \boldsymbol{v}_2, \ldots, \boldsymbol{A}_1, \boldsymbol{A}_2, \ldots, \boldsymbol{W}_1, \boldsymbol{W}_2, \ldots)$ of vectors $\boldsymbol{v}_\alpha(\alpha = 1, 2, \ldots, K)$, second-order symmetric tensors $\boldsymbol{A}_\beta(\beta = 1, 2, \ldots, L)$, and second-order skew-symmetric tensors $\boldsymbol{W}_\gamma$ $(\gamma = 1, 2, \ldots, N)$ is said to be an invariant of these vector and tensors under a given group of transformations $\boldsymbol{Q}$ if

$$f\left(\boldsymbol{v}_\alpha^*, \boldsymbol{A}_\beta^*, \boldsymbol{W}_\gamma^*\right) = (\det \boldsymbol{Q})^N f(\boldsymbol{v}_\alpha, \boldsymbol{A}_\beta, \boldsymbol{W}_\gamma) \tag{A.56}$$

for every transformation $\boldsymbol{Q}$ of the group. In the case $N = 0$, f is called an absolute invariant, and in the case $N \neq 0$, it is a relative invariant with weight N. We usually deal with absolute invariants.

Example 4. The scalar product of two vectors is an absolute invariant:

$$u_i^* v_i^* = Q_{ij} u_j Q_{il} v_l = \delta_{jl} u_j v_l = u_j v_j. \tag{A.57}$$

Example 5. The trace of a matrix is an absolute invariant:

$$A_{ii}^* = Q_{ik} Q_{il} A_{kl} = \delta_{kl} A_{kl} = A_{kk} = \operatorname{tr} \boldsymbol{A}. \tag{A.58}$$

Example 6. The determinant of matrix $\boldsymbol{A}$ is an absolute invariant

$$\det(A_{ij}^*) = \det(Q_{ik} Q_{il} A_{kl}) = (\det \boldsymbol{Q})^2 \det(A_{kl}) = \det(A). \tag{A.59}$$

If Eq. (A.56) with $N = 0$ is valid for all members of the full orthogonal group $\boldsymbol{Q}$, then the invariant function f is called *isotropic*. If it includes only the members of the full orthogonal group for which $\det \boldsymbol{Q} = +1$, we say that f is *hemitropic*. In this case, reflection is not allowed. If the function f in Eq. (A.56) is a polynomial in the components of vectors and tensors, then it is said to be a *polynomial invariant*. A set of invariants that can be used to express any invariant in members of the given set as a polynomial is called an "integrity" (a function) basis. A basis that contains the smallest possible number of members is called a "minimal" basis. The main problem of invariant theory is to determine the minimal basis.

Constitutive equations for vector $\boldsymbol{f}$-valued and tensor $\boldsymbol{T}$-valued functions require the invariance of the forms

$$\boldsymbol{f}(\boldsymbol{v}_\alpha^*, \boldsymbol{A}_\beta^*, \boldsymbol{W}_\gamma^*) = \boldsymbol{Q}\boldsymbol{f}(\boldsymbol{v}_\alpha, \boldsymbol{A}_\beta, \boldsymbol{W}_\gamma), \tag{A.60}$$

and

$$\boldsymbol{T}(\boldsymbol{v}_\alpha^*, \boldsymbol{A}_\beta^*, \boldsymbol{W}_\gamma^*) = \boldsymbol{Q}\boldsymbol{T}(\boldsymbol{v}_\alpha, \boldsymbol{A}_\beta, \boldsymbol{W}_\gamma)\boldsymbol{Q}^t. \tag{A.61}$$

These functions can be generated by use of certain products of the argument vectors and tensors, with coefficients that are functions of the invariants of the argument vectors and tensors. These vector and tensor products are called the generators of $\boldsymbol{f}$ and $\boldsymbol{T}$. For example, $\boldsymbol{T} = \boldsymbol{T}(\boldsymbol{A})$ (where $\boldsymbol{T}$ and $\boldsymbol{A}$ are symmetric), which obeys the invariance

$$\boldsymbol{T}(\boldsymbol{Q}\boldsymbol{A}\boldsymbol{Q}^T) = \boldsymbol{Q}\boldsymbol{T}(\boldsymbol{A})\boldsymbol{Q}^T$$

has the form

$$\boldsymbol{T} = a_0\boldsymbol{I} + a_1\boldsymbol{A} + a_2\boldsymbol{A}^2$$

where a_0, a_1, and a_2 are functions of invariants $\operatorname{tr} \boldsymbol{A}$, $\operatorname{tr} \boldsymbol{A}^2$, and $\operatorname{tr} \boldsymbol{A}^3$. The tensors $\boldsymbol{I}$, $\boldsymbol{A}$, and $\boldsymbol{A}^2$ are the generators of the tensor $\boldsymbol{T}$.

The representations for isotropic scalar, vector- and tensor-valued functions were studied by Wang (1969a,b, 1970, 1971), Smith (1970, 1971), and others. After the modifications discussed by Boehler (1977), both representations were made identical. They are summarized in the Appendix B.

Appendix B
Representations of Isotropic Scalar, Vector, and Tensor Functions (Wang, 1970, 1971)

TABLE B.1. Complete and irreducible sets of invariants of symmetric tensors $\boldsymbol{A}$, vectors $\boldsymbol{v}$, and skew-symmetric tensors $\boldsymbol{W}$.

Variables	Invariants
(I) Invariant depending on one variable	
$\boldsymbol{A}$	$\mathrm{tr}\boldsymbol{A}$, $\mathrm{tr}\boldsymbol{A}^2$, $\mathrm{tr}\boldsymbol{A}^3$
$\boldsymbol{v}$	$\boldsymbol{v}\cdot\boldsymbol{v}$
$\boldsymbol{W}$	$\mathrm{tr}\,\boldsymbol{W}^2$
(II) Invariant depending on two variables when (I) is assumed	
$\boldsymbol{A}_1, \boldsymbol{A}_2$	$\mathrm{tr}\boldsymbol{A}_1\boldsymbol{A}_2$, $\mathrm{tr}\boldsymbol{A}_1^2\boldsymbol{A}_2$, $\mathrm{tr}\boldsymbol{A}_1\,\boldsymbol{A}_2^2$, $\mathrm{tr}\boldsymbol{A}_1^1\boldsymbol{A}_2^2$
$\boldsymbol{A}, \boldsymbol{v}$	$\boldsymbol{v}\cdot\boldsymbol{A}\boldsymbol{v}$, $\boldsymbol{v}\cdot\boldsymbol{A}^2\boldsymbol{v}$
$\boldsymbol{A}, \boldsymbol{W}$	$\mathrm{tr}\boldsymbol{A}\boldsymbol{W}^2$, $\mathrm{tr}\boldsymbol{A}^2\boldsymbol{W}^2$, $\mathrm{tr}\boldsymbol{A}^2\boldsymbol{W}^2\boldsymbol{A}\boldsymbol{W}$
$\boldsymbol{v}_1, \boldsymbol{v}_2$	$\boldsymbol{v}_1\cdot\boldsymbol{v}_2$
$\boldsymbol{v}, \boldsymbol{W}$	$\boldsymbol{v}\cdot\boldsymbol{W}^2\boldsymbol{v}$
$\boldsymbol{W}_1, \boldsymbol{W}_2$	$\mathrm{tr}\,\boldsymbol{W}_1\boldsymbol{W}_2$
(III) Invariant depending on three variables when (II) is assumed	
$\boldsymbol{A}_1, \boldsymbol{A}_2, \boldsymbol{A}_3$	$\mathrm{tr}\boldsymbol{A}_1\,\boldsymbol{A}_2\,\boldsymbol{A}_3$
$\boldsymbol{A}_1, \boldsymbol{A}_2, \boldsymbol{v}$	$\boldsymbol{v}\cdot\boldsymbol{A}_1\,\boldsymbol{A}_2\boldsymbol{v}$
$\boldsymbol{A}, \boldsymbol{v}_1, \boldsymbol{v}_2$	$\boldsymbol{v}_1\cdot\boldsymbol{A}\boldsymbol{v}_2$, $\boldsymbol{v}_1\cdot\boldsymbol{A}^2\boldsymbol{v}_2$
$\boldsymbol{A}, \boldsymbol{W}_1, \boldsymbol{W}_2$	$\mathrm{tr}\boldsymbol{A}\boldsymbol{W}_1\boldsymbol{W}_2$, $\mathrm{tr}\boldsymbol{A}\boldsymbol{W}_1\boldsymbol{W}_2^2$, $\mathrm{tr}\boldsymbol{A}\boldsymbol{W}_1^2\boldsymbol{W}_2$
$\boldsymbol{A}_1, \boldsymbol{A}_2, \boldsymbol{W}$	$\mathrm{tr}\boldsymbol{A}_1\boldsymbol{A}_2\boldsymbol{W}$, $\mathrm{tr}\boldsymbol{A}_1^2\boldsymbol{A}_2\boldsymbol{W}$, $\mathrm{tr}\boldsymbol{A}_1\boldsymbol{A}_2^2\,\boldsymbol{W}$, $\mathrm{tr}\,\boldsymbol{A}_1\boldsymbol{W}^2\boldsymbol{A}_2$
$\boldsymbol{W}_1, \boldsymbol{W}_2, \boldsymbol{W}_3$	$\mathrm{tr}\,\boldsymbol{W}_1\,\boldsymbol{W}_2\,\boldsymbol{W}_3$
$\boldsymbol{v}_1, \boldsymbol{v}_2, \boldsymbol{W}$	$\boldsymbol{v}_1\cdot\boldsymbol{W}\boldsymbol{v}_2$, $\boldsymbol{v}_1\cdot\boldsymbol{W}^2\boldsymbol{v}_2$
$\boldsymbol{v}, \boldsymbol{W}_1, \boldsymbol{W}_2$	$\boldsymbol{v}\cdot\boldsymbol{W}_1\,\boldsymbol{W}_2\,\boldsymbol{v}$, $\boldsymbol{v}\cdot\boldsymbol{W}_1^2\,\boldsymbol{W}_2\,\boldsymbol{v}$, $\boldsymbol{v}\cdot\boldsymbol{W}_1\,\boldsymbol{W}_2^2\boldsymbol{v}$
$\boldsymbol{A}, \boldsymbol{v}, \boldsymbol{W}$	$\boldsymbol{v}\cdot\boldsymbol{A}\boldsymbol{W}\boldsymbol{v}$, $\boldsymbol{v}\cdot\boldsymbol{A}^2\,\boldsymbol{W}\boldsymbol{v}$, $\boldsymbol{v}\cdot\boldsymbol{A}\boldsymbol{W}^2\boldsymbol{v}$
(IV) Invariant depending on four variables when (III) is assumed	
$\boldsymbol{A}_1, \boldsymbol{A}_2, \boldsymbol{v}_1, \boldsymbol{v}_2$	$\boldsymbol{v}_1\cdot(\boldsymbol{A}_1\boldsymbol{A}_2 - \boldsymbol{A}_2\,\boldsymbol{A}_1)\boldsymbol{v}$
$\boldsymbol{A}, \boldsymbol{v}_1, \boldsymbol{v}_2, \boldsymbol{W}$	$\boldsymbol{v}_1\cdot(\boldsymbol{A}\boldsymbol{W} - \boldsymbol{W}\boldsymbol{A})\boldsymbol{v}_2$

TABLE B.2. Generators for vector-valued isotropic functions.

Variables	Generator
(I) Generators depending on one variable	
$\boldsymbol{v}$	$\boldsymbol{v}$
(II) Generators depending on two variables when (I) is assumed	
$\mathbf{A}, \boldsymbol{v}$	$\boldsymbol{A}\boldsymbol{v}, \boldsymbol{A}^2\boldsymbol{v}$
$\mathbf{W}, \boldsymbol{v}$	$\boldsymbol{W}\boldsymbol{v}, \boldsymbol{W}^2\boldsymbol{v}$
(III) Generators depending on three variables when (II) is assumed	
$\boldsymbol{A}_1, \boldsymbol{A}_2, \boldsymbol{v}$	$(\boldsymbol{A}_1 \boldsymbol{A}_2 - \boldsymbol{A}_2 \boldsymbol{A}_1)\boldsymbol{v}$
$\boldsymbol{W}_1, \boldsymbol{W}_2, \boldsymbol{v}$	$(\boldsymbol{W}_1 \boldsymbol{W}_2 - \boldsymbol{W}_2 \boldsymbol{W}_1)\boldsymbol{v}$
$\boldsymbol{A}, \boldsymbol{v}, \boldsymbol{W}$	$(\boldsymbol{A}\boldsymbol{W} - \boldsymbol{W}\boldsymbol{A})\boldsymbol{v}$

TABLE B.3. Generators for symmetric tensor-valued isotropic functions.

Variables	Generator
(I) Generators depending on no variable	
	$\boldsymbol{I}$
(II) Generators depending on one variable when (I) is assumed	
$\boldsymbol{A}$	$\boldsymbol{A}, \boldsymbol{A}^2$
$\boldsymbol{v}$	$\boldsymbol{v} \otimes \boldsymbol{v}$
$\boldsymbol{W}$	$\boldsymbol{W}^2$
(III) Generators depending on two variables when (II) is assumed	
$\boldsymbol{A}_1, \boldsymbol{A}_2$	$\boldsymbol{A}_1 \boldsymbol{A}_2 + \boldsymbol{A}_2 \boldsymbol{A}_1, \boldsymbol{A}_1^2 \boldsymbol{A}_2 + \boldsymbol{A}_2 \boldsymbol{A}_1^2, \boldsymbol{A}_1 \boldsymbol{A}_2^2 + \boldsymbol{A}_2^2 \boldsymbol{A}_1$
$\boldsymbol{A}, \boldsymbol{v}$	$\boldsymbol{v} \otimes \boldsymbol{A}\boldsymbol{v} + \boldsymbol{A}\boldsymbol{v} \otimes \boldsymbol{v}, \boldsymbol{v} \otimes \boldsymbol{A}^2\boldsymbol{v} + \boldsymbol{A}^2\boldsymbol{v} \otimes \boldsymbol{v}$
$\boldsymbol{A}, \boldsymbol{W}$	$\boldsymbol{A}\boldsymbol{W} - \boldsymbol{W}\boldsymbol{A}, \boldsymbol{W}\boldsymbol{A}\boldsymbol{W}, \boldsymbol{A}^2\boldsymbol{W} - \boldsymbol{W}\boldsymbol{A}^2, \boldsymbol{W}\boldsymbol{A}\boldsymbol{W}^2 - \boldsymbol{W}^2\boldsymbol{A}\boldsymbol{W}$
$\boldsymbol{v}_1, \boldsymbol{v}_2$	$\boldsymbol{v}_1 \otimes \boldsymbol{v}_2 + \boldsymbol{v}_2 \otimes \boldsymbol{v}_1$
$\boldsymbol{v}, \boldsymbol{W}$	$\boldsymbol{W}\boldsymbol{v} \otimes \boldsymbol{W}\boldsymbol{v}, \boldsymbol{v} \otimes \boldsymbol{W}\boldsymbol{v} + \boldsymbol{W}\boldsymbol{v} \otimes \boldsymbol{v}, \boldsymbol{W}\boldsymbol{v} \otimes \boldsymbol{W}^2\boldsymbol{v} + \boldsymbol{W}^2\boldsymbol{v} \otimes \boldsymbol{W}\boldsymbol{v}$
$\boldsymbol{W}_1, \boldsymbol{W}_2$	$\boldsymbol{W}_1 \boldsymbol{W}_2 + \boldsymbol{W}_2 \boldsymbol{W}_1, \boldsymbol{W}_1 \boldsymbol{W}_2^2 - \boldsymbol{W}_2^2 \boldsymbol{W}_1, \boldsymbol{W}_1^2 \boldsymbol{W}_2 - \boldsymbol{W}_2 \boldsymbol{W}_1^2$
(IV) Generators depending on three variables, (III) is assumed	
$\boldsymbol{A}, \boldsymbol{v}_1, \boldsymbol{v}_2$	$\boldsymbol{A}(\boldsymbol{v}_1 \otimes \boldsymbol{v}_2 - \boldsymbol{v}_2 \otimes \boldsymbol{v}_2) - (\boldsymbol{v}_1 \otimes \boldsymbol{v}_2 - \boldsymbol{v}_2 \otimes \boldsymbol{v}_1)\boldsymbol{A}$
$\boldsymbol{W}, \boldsymbol{v}_1, \boldsymbol{v}_2$	$\boldsymbol{W}(\boldsymbol{v}_1 \otimes \boldsymbol{v}_2 - \boldsymbol{v}_2 \otimes \boldsymbol{v}_2) + (\boldsymbol{v}_1 \otimes \boldsymbol{v}_2 - \boldsymbol{v}_2 \otimes \boldsymbol{v}_1)\boldsymbol{W}$

TABLE B.4. Generators for skew-symmetric tensor-valued isotropic functions.

Variables	Generator
(I) Generators depending on one variable	
$\boldsymbol{W}$	$\boldsymbol{W}$
(II) Generators depending on two variables when (I) is assumed	
$\boldsymbol{A}_1, \boldsymbol{A}_2$	$\boldsymbol{A}_1\boldsymbol{A}_2 - \boldsymbol{A}_2\boldsymbol{A}_1, \boldsymbol{A}_1^2\boldsymbol{A}_2 - \boldsymbol{A}_2\boldsymbol{A}_1^2, \boldsymbol{A}_1\boldsymbol{A}_2^2 - \boldsymbol{A}_2^2\boldsymbol{A}_1,$ $\boldsymbol{A}_1\boldsymbol{A}_2\boldsymbol{A}_1^2 - \boldsymbol{A}_1^2\boldsymbol{A}_2\boldsymbol{A}_1, \boldsymbol{A}_2\boldsymbol{A}_1\boldsymbol{A}_2^2 - \boldsymbol{A}_2^2\boldsymbol{A}_1\boldsymbol{A}_2$
$\boldsymbol{A}, \boldsymbol{v}$	$\boldsymbol{v}\otimes\boldsymbol{A}\boldsymbol{v} - \boldsymbol{A}\boldsymbol{v}\otimes\boldsymbol{v}, \boldsymbol{v}\otimes\boldsymbol{A}^2\boldsymbol{v} - \boldsymbol{A}^2\boldsymbol{v}\otimes\boldsymbol{v},$ $\boldsymbol{A}\boldsymbol{v}\otimes\boldsymbol{A}^2\boldsymbol{v} - \boldsymbol{A}^2\boldsymbol{v}\otimes\boldsymbol{A}\boldsymbol{v}$
$\boldsymbol{A}, \boldsymbol{W}$	$\boldsymbol{A}\boldsymbol{W} + \boldsymbol{W}\boldsymbol{A}, \boldsymbol{A}\boldsymbol{W}^2 - \boldsymbol{W}^2\boldsymbol{A}$
$\boldsymbol{W}, \boldsymbol{v}$	$\boldsymbol{v}\otimes\boldsymbol{W}\boldsymbol{v} - \boldsymbol{W}\boldsymbol{v}\otimes\boldsymbol{v}, \boldsymbol{v}\otimes\boldsymbol{W}^2\boldsymbol{v} - \boldsymbol{W}^2\boldsymbol{v}\otimes\boldsymbol{v}$
$\boldsymbol{v}_1, \boldsymbol{v}_2$	$\boldsymbol{v}_1\otimes\boldsymbol{v}_2 - \boldsymbol{v}_2\otimes\boldsymbol{v}_1$
$\boldsymbol{W}_1, \boldsymbol{W}_2$	$\boldsymbol{W}_1\boldsymbol{W}_2 - \boldsymbol{W}_2\boldsymbol{W}_1^2$
(III) Generators depending on three variables when (II) is assumed	
$\boldsymbol{A}_1, \boldsymbol{A}_2, \boldsymbol{A}_3$	$\boldsymbol{A}_1\boldsymbol{A}_2\boldsymbol{A}_3 + \boldsymbol{A}_2\boldsymbol{A}_3\boldsymbol{A}_1 + \boldsymbol{A}_3\boldsymbol{A}_1\boldsymbol{A}_2$ $- \boldsymbol{A}_2\boldsymbol{A}_1\boldsymbol{A}_3 - \boldsymbol{A}_1\boldsymbol{A}_3\boldsymbol{A}_2 - \boldsymbol{A}_3\boldsymbol{A}_2\boldsymbol{A}_1$
$\boldsymbol{A}_1, \boldsymbol{A}_2, \boldsymbol{v}$	$\boldsymbol{A}_1\boldsymbol{v}\otimes\boldsymbol{A}_2\boldsymbol{v} - \boldsymbol{A}_2\boldsymbol{v}\otimes\boldsymbol{A}_1\boldsymbol{v},$ $\boldsymbol{v}\otimes(\boldsymbol{A}_1\boldsymbol{A}_2 - \boldsymbol{A}_2\boldsymbol{A}_1)\boldsymbol{v} - (\boldsymbol{A}_1\boldsymbol{A}_2 - \boldsymbol{A}_2\boldsymbol{A}_1)\boldsymbol{v}\otimes\boldsymbol{v}$
$\boldsymbol{A}, \boldsymbol{v}_1, \boldsymbol{v}_2$	$\boldsymbol{A}(\boldsymbol{v}_1\otimes\boldsymbol{v}_2 - \boldsymbol{v}_2\otimes\boldsymbol{v}_2) + (\boldsymbol{v}_1\otimes\boldsymbol{v}_2 - \boldsymbol{v}_2\otimes\boldsymbol{v}_1)\boldsymbol{A}$
$\boldsymbol{W}, \boldsymbol{v}_1, \boldsymbol{v}_2$	$\boldsymbol{W}(\boldsymbol{v}_1\otimes\boldsymbol{v}_2 - \boldsymbol{v}_2\otimes\boldsymbol{v}_2) - (\boldsymbol{v}_1\otimes\boldsymbol{v}_2 - \boldsymbol{v}_2\otimes\boldsymbol{v}_1)\boldsymbol{W}$

Appendix C
Classification of Partial Differential Equations

In the theory and application of ordinary or partial differential equations, the dependent variable, denoted by u, is usually required to satisfy some conditions on the boundary of the domain on which the differential equation is defined. The equations that represent those boundary conditions may involve values of derivatives of u, as well as values of u itself, at points on the boundary. In addition, some conditions on the continuity of u and its derivatives within the domain and on the boundary may be required.

Such a set of requirements constitutes a *boundary value problem*. A boundary value problem is correctly set if it has one and only one solution within a given class of functions. Physical interpretations often suggest boundary conditions under which a problem may be correctly set. In fact, it is sometimes helpful to interpret a problem physically in order to judge whether the boundary conditions may be adequate. This is a prominent reason for associating such problems with their physical applications, aside from the opportunity to illustrate connections between mathematical analysis and the physical sciences.

The theory of partial differential equations gives results on the existence and uniqueness of solutions of boundary value problems. A *partial differential equation* is an equation that involves an unknown function and some of its derivatives with respect to two or more independent variables. An nth order equation has its highest order derivative of order n. A partial differential equation is *linear* if it is an equation of the first degree in the dependent variable and its derivatives. A partial differential equation is *homogeneous* if every term contains the dependent variable or one of its partial derivatives. The trivial (zero) function is always a solution of a homogeneous equation.

The general linear partial differential equation of the second order in $u = u(x, y)$ has the form

$$Au_{xx} + Bu_{xy} + Cu_{yy} + Du_x + Eu_y + Fu = G, \tag{C.1}$$

where the coefficients are functions of the independent variables x and y, and we use the subscript notation to denote partial derivatives:

$$u_x \text{ or } u_x(x, y) \text{ for } \frac{\partial u}{\partial x}, \quad u_{xx} \text{ for } \frac{\partial^2 u}{\partial x^2}, \quad u_{xy} \text{ for } \frac{\partial^2 u}{\partial x \partial y}, \ldots \tag{C.2}$$

We shall always assume that the partial derivatives of u satisfy conditions allowing us to write $u_{xy} = u_{yx}$.

Example 1. Linear and nonlinear partial differential equations.
The differential equation

$$xu_{xx} + y^2u_{xy} + xyu_{yy} + xy + x = 1 \tag{C.3}$$

is linear in $u = u(x, y)$, but the equation

$$u_{xx} + uu_y = 1 \tag{C.4}$$

is nonlinear in $u = u(x, y)$ because the term uu_y is not of first degree as an algebraic expression in the two variables u and u_y.

Heat Equation

Thermal energy is transferred from warmer to cooler regions interior to a solid body by means of conduction. It is convenient to refer to the transfer as a flow of heat, as if heat were a fluid or gas that diffused through the body from regions of high concentration into regions of low concentration. A fundamental postulate in the mathematical theory of heat conduction is *Fourier's law.* If $T(x, y, z, t)$ denotes temperatures at points of body at time t, n a direction of $\boldsymbol{n}$, $\boldsymbol{q}$ the rate of heat flow, Fourier's law states that

$$\boldsymbol{q} = -k\frac{\mathrm{d}T}{\mathrm{d}\boldsymbol{n}} = -k\frac{\partial T}{\partial x}\boldsymbol{e}_x - k\frac{\partial T}{\partial y}\boldsymbol{e}_y - k\frac{\partial T}{\partial z}\boldsymbol{e}_z, \tag{C.5}$$

or

$$\boldsymbol{q} = -k\nabla T. \tag{C.6}$$

Here, the material isotropy has been assumed and so the same conductivity k for three directions can be used. The heat flux $\boldsymbol{q}$ has the units of energy per unit area per unit time.

Suppose now an arbitrary body of volume V, closed surface S, and unit outward normal $\boldsymbol{n}$ (Fig C.1). The balance of energy for this body is that the heat entering the body through the surface per unit time plus the heat produced by sources in the body per unit time equals the time rate of change of the heat content of the

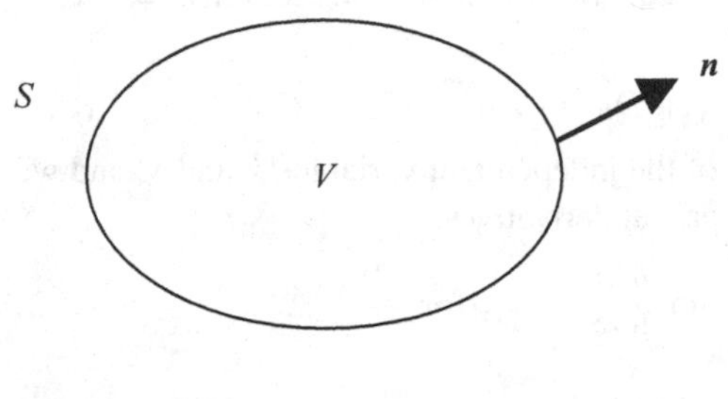

FIGURE C.1. Arbitrary body of volume V and surface S

body:

$$\oint_S -\boldsymbol{q}\cdot\boldsymbol{n}\,\mathrm{d}S + \int_V Q\,\mathrm{d}V = \int_V \rho c \frac{\partial T}{\partial t}\,\mathrm{d}V, \tag{C.7}$$

where $Q = Q(x, y, z)$ is the internal heat generation per unit volume per unit time, ρ the mass density of the material, c the specific heat (the heat required per unit mass to raise the temperature by 1°), and t time. Substituting Fourier's law of the heat conduction, Eq. (C.6), into Eq. (C.7), and applying the divergence theorem, we have

$$\int_V \left\{ \nabla \cdot k(\nabla T) + Q - \rho c \frac{\partial T}{\partial t} \right\} \mathrm{d}V = 0. \tag{C.8}$$

Since this equation holds for arbitrary volume, this requires

$$\nabla \cdot (k \nabla T) + Q - \rho c \frac{\partial T}{\partial t} = 0 \text{ in } V, \tag{C.9}$$

or

$$\frac{\partial}{\partial x}\left(k\frac{\partial T}{\partial x}\right) + \frac{\partial}{\partial y}\left(k\frac{\partial T}{\partial y}\right) + \frac{\partial}{\partial z}\left(k\frac{\partial T}{\partial z}\right) + Q = \rho c \frac{\partial T}{\partial t}. \tag{C.10}$$

This is the famous *heat equation*. It is also called the *diffusion equation*. If k is independent of the space variables (x, y, z), which is referred as a homogeneous material, the heat equation becomes

$$\nabla^2 T = \frac{1}{K}\frac{\partial T}{\partial t} - \frac{Q}{k}, \tag{C.11}$$

where $K = k/(\rho c)$ is the *thermal diffusivity*. The steady-state temperature in a system with a homogeneous material satisfies *Poisson's equation*:

$$\nabla^2 T = -Q/k. \tag{C.12}$$

In the absence of internal source, $Q = 0$, it results *Laplace's equation*:

$$\nabla^2 T = 0. \tag{C.13}$$

A function that is continuous, together with its partial derivatives of the first order and second order, and satisfies Laplace's equation is called a *harmonic function*.

Wave Equation

Consider a tightly stretched string, whose position of equilibrium is some interval on the x-axis, vibrating in the xy plane. Each point of the string, with coordinates $(x, 0)$ in the equilibrium position, has a transverse displacement $u = u(x, t)$ at time t. We assume that both the displacements $u(x, y)$ and the slopes $\partial u/\partial x$ are very small and that the movement of each point is vertical to the x-axis. Denoting the string tension as T at location x, the tension at short distance $\mathrm{d}x$ away can be obtained using Taylor's series approximation

$$T + \mathrm{d}T \approx T + \frac{\partial T}{\partial x}\mathrm{d}x. \tag{C.14}$$

Similarly given the slope $\partial u/\partial x$ at x, the slope at $x + \mathrm{d}x$ is approximately

$$\frac{\partial u}{\partial x} + \frac{\partial}{\partial x}\left(\frac{\partial u}{\partial x}\right)\mathrm{d}x = \frac{\partial u}{\partial x} + \frac{\partial^2 u}{\partial x^2}\mathrm{d}x. \tag{C.15}$$

Applying Newton's second law, the net applied force is written as

$$\left(T + \frac{\partial T}{\partial x}\mathrm{d}x\right)\left(\frac{\partial u}{\partial x} + \frac{\partial^2 u}{\partial x^2}\mathrm{d}x\right) - T\frac{\partial u}{\partial x} + f(x,t) = \rho A\,\mathrm{d}x\frac{\partial^2 u}{\partial x^2}, \tag{C.16}$$

where ρ is the density of the string material and A the cross-sectional area. Eliminating the nonlinear terms, we obtain the linear partial differential equation for the vibrating string

$$T\frac{\partial^2 u}{\partial x^2} + f(x,t) = \rho A\frac{\partial^2 u}{\partial t^2}, \tag{C.17}$$

or

$$u_{xx} + \frac{f(x,t)}{T} = \frac{1}{c^2}u_{tt}, \tag{C.18}$$

where $c = \sqrt{T/\rho A}$.

For zero force ($f = 0$), Eq. (C.18) reduces to

$$u_{xx} = u_{tt}/c^2. \tag{C.19}$$

This is the one-dimensional wave equation. The transverse displacement for the unforced, infinitesimal vibrations of a stretched string satisfies the one-dimensional wave equation. The transverse displacement for the unforced, infinitesimal vibrations of membranes satisfies the two-dimensional wave equation

$$u_{xx} + u_{yy} = u_{tt}/c^2. \tag{C.20}$$

In three dimensions, the wave equation becomes

$$u_{xx} + u_{yy} + u_{zz} = u_{tt}/c^2, \tag{C.21}$$

which is the applicable equation for acoustics, where the variable u represents, for example, the pressure p or the velocity potential ϕ.

In any number of dimensions, the wave equation can therefore be written as

$$\nabla^2 u = \frac{1}{c^2}u_{tt}. \tag{C.22}$$

For time-harmonic motion, $u = u_0\cos(\omega t)$, the wave equation then reduces to the *Helmholtz equation*,

$$\nabla^2 u_0 + k^2 u_0 = 0, \tag{C.23}$$

where $k = \omega/c$ is called the wave number. The Helmholtz equation is sometimes referred to as the *reduced wave equation.*

Classification of Partial Differential Equations

The second-order linear partial differential equation

$$Au_{xx} + Bu_{xy} + Cu_{yy} + Du_x + Eu_y + Fu = G, \tag{C.24}$$

in $u = u(x, y)$, where A, B, C, D, E, F, and G are constants or functions of x and y, is classified in any given region of the xy plane according to whether $B^2 - 4AC$ is positive, negative, or zero throughout that region. Specifically, Eq. (C.24) is

$$\begin{aligned} &\text{(a) Hyperbolic if } B^2 - 4AC > 0;\\ &\text{(b) Elliptic if } B^2 - 4AC < 0;\\ &\text{(c) Parabolic if } B^2 - 4AC = 0. \end{aligned} \tag{C.25}$$

For each of these categories, Eq. (C.24) and its solutions have distinct features, and the behavior of the solutions will differ. Elliptic equations characterize static problems and the hyperbolic or parabolic equations characterize time-dependent problems.

Example 2. Laplace equation in two dimensions

$$u_{xx} + u_{yy} = 0 \tag{C.26}$$

is a special case of Eq. (C.24) in which $A = C = 1$ and $B = 0$, $B^2 - 4AC < 0$. Hence, it is elliptic throughout xy plane. It arises in incompressible fluid flow, gravitational potential problems, electrostatics, magnetostatics, and steady-state heat conduction.

Example 3. Poisson's equation in two dimensions

$$u_{xx} + u_{yy} = f(x, y) \tag{C.27}$$

has $A = C = 1$ and $B = 0$. It is elliptic in any region of xy plane, where $f(x, y)$ is defined. It may appear in steady-state heat conduction with distributed sources and torsion of prismatic bars in elasticity.

Example 4. Helmholtz equation

$$u_{xx} + u_{yy} + k^2u = 0 \tag{C.28}$$

is also elliptic in two dimensions. It arises in time-harmonic elastic vibrations (strings, bars, and membranes), acoustics, and electromagnetics.

Example 5. The one-dimensional heat equation

$$-ku_{xx} + u_t = 0 \tag{C.29}$$

is parabolic in the xt plane. It arises in heat conduction and other diffusion processes.

Example 6. The one-dimensional wave equation

$$-c^2 u_{xx} + u_{tt} = 0 \tag{C.30}$$

is hyperbolic in $u = u(x, t)$. It arises in transient problems.

The type of partial differential equations is of special importance in dealing with localization problems. From a mathematical point of view, the appearance of localization in classical local continuum mechanics with rate independence is associated with the change of type of the differential governing equations: loss of ellipticity in quasi-static problems; change from hyperbolic to elliptic type in the dynamic case.

Type of Boundary Conditions and Uniqueness of Solution

The three types of second-order linear equations require different types of boundary conditions in order to determine a solution.

Let u denote the dependent variable in a boundary value problem. A condition that prescribes the value of u itself along a portion of the boundary is known as a *Dirichlet condition*, or *essential boundary condition*. A condition that prescribes the value of $\mathrm{d}u/\mathrm{d}n$ on a part of boundary is called *Neumann condition*, or *natural boundary condition*, or *nonessential boundary condition*. A linear combination of both is a *Robin condition*. It prescribes the value of $hu + \mathrm{d}u/\mathrm{d}n$ at boundary points, where h is either a constant or a function of the independent variables.

A *well-posed* problem is one for which the solution exists, is unique, and has continuous dependence on the data. For transient heat conduction problems, the partial differential equation

$$\nabla^2 T = \frac{1}{K}\frac{\partial T}{\partial t} - \frac{Q}{k} \text{ in } V \tag{C.31}$$

has a unique solution $T(x, y, z, t)$ if there are imposed boundary conditions

$$\alpha \frac{\partial T}{\partial n} + bT = f(x, y, z, t) \text{ on } S, \tag{C.32}$$

where a and b are constants, and initial condition

$$T(x, y, z, 0) = T_0(x, y, z). \tag{C.33}$$

For general linear partial differential equations, a nonhomogeneous system has a unique solution if and only if the corresponding homogeneous system has only the trivial solution.

For steady-state (time-independent) heat transfer problem, for which the relevant differential equation is the Poisson or Laplace equation, the general requirement for uniqueness is that a boundary condition of the form

$$\alpha \frac{\partial T}{\partial n} + bT = f(x, y, z, t) \tag{C.34}$$

be imposed.

A contrary example is a pure Neumann problem

$$\begin{cases} \nabla^2 T = f \text{ in } V, \\ \dfrac{\partial T}{\partial n} = g \text{ on } S. \end{cases} \tag{C.35}$$

Uniqueness for this nonhomogeneous problem requires that $T = 0$ is the only solution of the corresponding homogeneous problem

$$\begin{cases} \nabla^2 T = 0 \text{ in } V, \\ \dfrac{\partial T}{\partial n} = 0 \text{ on } S. \end{cases} \tag{C.36}$$

However, since every appearance of T in the above system is in a derivative, solutions of Neumann problems are unique only up to an arbitrary additive constant. Guaranteeing uniqueness in such cases requires that T be specific at one point.

Nonuniqueness in this Neumann problem is analogous to the nonuniqueness that occurs in the static mechanical problem in which a free bar is loaded in tension with a force F at each end. Although the stresses in the bar are unique, the displacements are not unique unless the location of one point is prescribed.

Appendix D
User's Manual: Meshless Computer Program for Electrostatics and Electrodynamics

Meshless Analysis of Elasticity

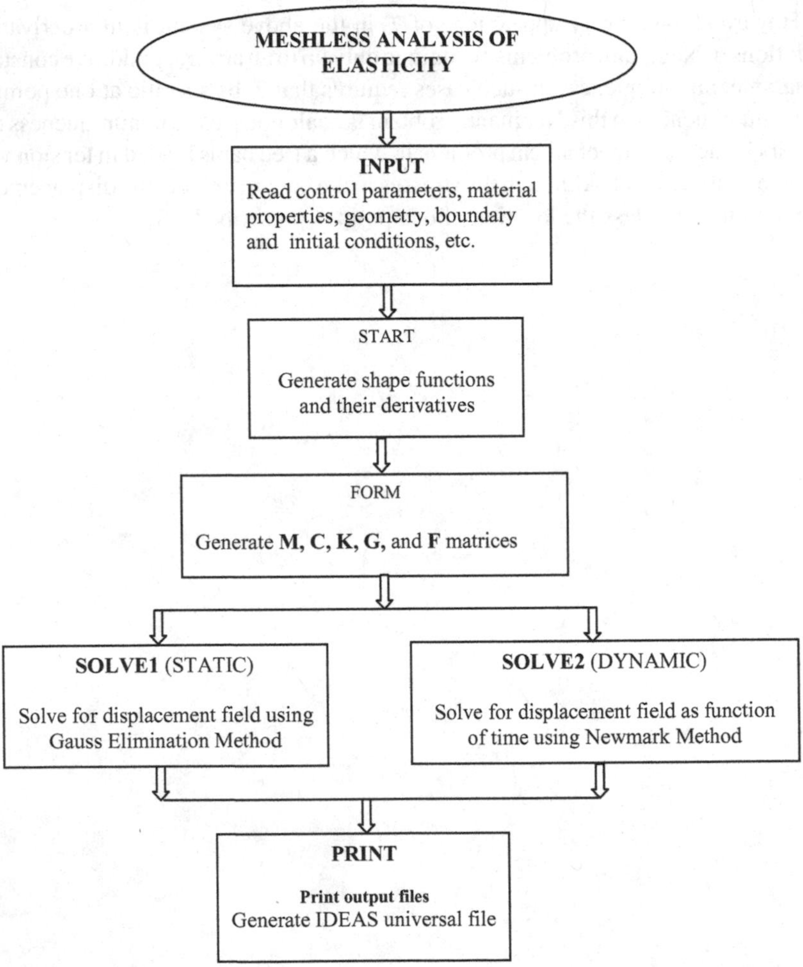

TABLE D.1. Table of input data (INFILE).

Variables	Format	Section
ID(1), ID(2), ..., ID(7)	16I5	1
MP, ME	9X, 2I8	2
1, (IJK(J,1), J = 1,4)	9X, I8, 8X, 4I8	
2, (IJK(J,2), J = 1,4)		
⋮		
⋮		
ME, (IJK(J,ME), J = 1,4)	9X, I8, 7X, 2G8.5	
1, XMESH(1), YMESH(1)		
2, XMESH(2), YMESH(2)		
⋮		
⋮		
MP, XMESH(MP), YMESH(MP)		
MN	9X, I8	3
1, XNODE(1), YNODE(1), RNODE(1)	9X, I8, 7X, 3G8.5	
2, XNODE(2), YNODE(2), RNODE(2)		
⋮		
⋮		
MN, XNODE(MN), YNODE(MN), RNODE(MN)		
DEN, RADIUS	2E15.8	4
YOUNG, POISSON		
YDAMP, PDAM		
MU, MF	2I5	5
ICU(1), XB(1), YB(1), SCALE1(1)	I5, 3(5X, F15.8)	6
ICU(2), XB(2), YB(2), SCALE1(2)		
⋮		
⋮		
ICU(MU), XB(MU), YB(MU), SCALE1(MU)		
ICF(1), SCALE2(1)	I5, 5X, F15.8	7
ICF(2), SCALE2(2)		
⋮		
⋮		
ICF(MF), SCALE2(MF)		
DTIME, TREF	2E15.8	8
XBARRIER(1), YBARRIER(1)	2E15.8	9
XBARRIER(2), YBARRIER(2)		
MUPRINT, MSPRINT	2I5	10
(IUPRINT(K), K = 1, MUPRINT)	4I5	
(ISPRINT(L), L = 1, MSPRINT)	3I5	

Control Parameters

Definitions

ID(1) = 1	Static case
ID(1) = 2	Dynamic case (Newmark method)
ID(2) = 1	Local theory
ID(2) = 2	Nonlocal theory
ID(3) = 0	(For further development)
ID(4) = 1	Plane strain
ID(4) = 2	Plane stress
ID(5) = 0	There is no barrier
ID(5) = 1	There is a barrier
ID(6) = MTIME	Number of time steps in the dynamic analysis
ID(7) = IPRINT	Outputs, including IDEAS, at every IPRINT time step

Remarks

1. This is a general purpose computer program based on local and nonlocal continuum field theory and meshless method.
2. This version is focused on two-dimensional elasticity, including static and dynamic analyses.
3. The output includes an IDEAS universal file that can be used to graphically display stress distributions of the whole specimen.

Background Mesh

Definitions

MP	Number of points in the background mesh
ME	Number of elements in the background mesh
IJK(J,I)	The number of the *J*th point of the *I*th element
XMESH(I)	The *X*-coordinate of the *I*th point in the background mesh
YMESH(I)	The *Y*-coordinate of the *I*th point in the background mesh

Remarks

1. The background mesh is like a finite element mesh. It is used for two purposes: (1) to generate an IDEAS universal file and (2) to create a set of sampling points, each of which is associated with an area.
2. If the graphic display is no longer a concern, then we do not need the background mesh; instead, we need a set of sampling points, each of which is associated with an area. The sampling points are equivalent to the Gauss points in finite element analysis.

Nodes

Definitions

MN	Number of nodes
XNODE(I)	The *X*-coordinate of the *I*th node
YNODE(I)	The *Y*-coordinate of the *I*th node
RNODE(I)	The radius of support of the *I*th node

Remarks

The weight function for a generic sampling point $\mathbf{x}$ and a node $\mathbf{x}_I$ is a quartic spline expressed as

$$w(\mathbf{x}, \mathbf{x}_I) \equiv w_I(\mathbf{x}) = \begin{cases} 1 - 6s^2 + 8s^3 - 3s^4, & \text{if } s \leq 1 \\ 0, & \text{if } s \geq 1 \end{cases}$$

where R_I is the radius of support of the *I*th node and

$$s \equiv \frac{||\mathbf{x} - \mathbf{x}_I||}{R_I}.$$

Material Properties

Definitions

DEN	The density of the material	ρ
RADIUS	The radius of nonlocality (nonlocal theory)	R
YOUNG	Young's modulus of the material	E
POISSON	The Poisson's ration of the material	υ
YDAMP	The first damping coefficient of the material (corresponding to Young's modulus in elasticity)	$\hat{E}$
PDAMP	The second damping coefficient of the material $\hat{\upsilon}$ (corresponding to Poisson's ratio in elasticity)	

Remarks

The constitutive relation is based on the Kelvin–Voigt model, in the local theory it is expressed as

$$\sigma_{ij}(\mathbf{x}) = \lambda\varepsilon_{kk}(\mathbf{x})\delta_{ij} + 2\mu\varepsilon_{ij}(\mathbf{x}) + \hat{\lambda}\dot{\varepsilon}_{kk}(\mathbf{x})\delta_{ij} + 2\hat{\mu}\dot{\varepsilon}_{ij}(\mathbf{x}), \qquad \text{(D.4.1)}$$

where Young's modulus and Poisson's ratio are related to the two Lame constants λ and μ; in the same manner, YDAMP and PDAMP are related to $\hat{\lambda}$ and $\hat{\mu}$. Note that the equation above emphasizes that the stresses at a point are related to the strains and strain rates at that point only.

For nonlocal theory, the constitutive relation is generalized to be

$$\sigma_{ij}(\mathbf{x}) = C(\mathbf{x})\left\{A_{ijmn}\int_{\Omega'} f(r)\varepsilon_{mn}(\mathbf{x}')d\Omega(\mathbf{x}') + a_{ijmn}\int_{\Omega'} f(r)\dot{\varepsilon}_{mn}(\mathbf{x}')d\Omega(\mathbf{x}')\right\}, \tag{D.4.2}$$

where

$$C(\mathbf{x}) = \frac{1}{\int_{\Omega'} f(r)\mathrm{d}\Omega(\mathbf{x}')}$$

$$f(r) = \begin{cases} 1 - 6r^2 + 8r^3 - 3r^4, & \text{if } r \le 1 \\ 0, & \text{if } r \ge 1 \end{cases}$$

$$r \equiv \frac{||\mathbf{x} + \mathbf{x}'||}{R}$$

and R is the radius of nonlocality. Note that as $R \to 0$, $f(r)$ behaves like the δ function and then Eq. (D.4.2) is reduced to Eq. (D.4.1).

Boundary Condition Parameters

Definitions

MU	Number of displacement-specified boundary conditions (see Section D.6)
MF	Number of nodal-force-specified boundary conditions (see Section D.7)

Displacement-Specified Boundary Condition

Definitions

ICU(I)	The *X*- or *Y*-component of the *I*th displacement-specified boundary condition
XB(I)	The *X*-coordinate of the location at which the *I*th boundary condition is specified
YB(I)	The *Y*-coordinate of the location at which the *I*th boundary condition is specified
SCALE1(I)	The magnitude of the displacement specified for the ICU(I)th component

Remarks

For example, the fifth displacement-specified boundary condition is that the Y-component of the displacement at $x = 1.23$, $y = 4.56$ is equal to 7.89, then we specify the following

$$\mathrm{ICU}(5) = 2$$
$$\mathrm{XB}(5) = 1.23$$
$$\mathrm{YB}(5) = 4.56$$
$$\mathrm{SCALE1}(5) = 7.89$$

Note that $\{x, y\} = \{1.23, 4.56\}$ has to be on the boundary but does not have to be a node.

In dynamic analysis, this boundary condition is read as

$$u_y(x_B, y_B, t) = 7.89 \times f(t)$$

where $f(t)$ is the time factor (cf. Section D.8).

Nodal Force Specified Boundary Condition

Definitions

ICF(I)	The component number of the Ith nodal force specified boundary condition
SCALE2(I)	The magnitude of the nodal force specified for the ICF(I)-th component

Remarks

The nodal-force-specified boundary condition is a type of natural boundary condition. There is no difference between finite element method and meshless method. For example, if the first natural boundary condition is that the Y-component of the applied force at the 43rd node equals 64.126, then we set

$$\mathrm{ICF}(1) = 86$$
$$\mathrm{SCALE2}(1) = 64.126$$

In dynamic analysis, this boundary condition is read as

$$F_y(43) = 64.126 \times f(t)$$

where $f(t)$ is the time factor (cf. Section D.8).

Dynamic Analysis Parameters

Definitions

DTIME	The magnitude of the time step used in the dynamic analysis
TREF	The reference time used in the user supplied subroutine FTIME to define the dynamic loading

Remarks

DTIME is the Δt used in the dynamic analysis. In static analysis, the value of DTIME is irrelevant.

The time factor $f(t)$ is defined in a user-supplied subroutine, named FTIME, in most cases, $f(t)$ reduces to zero after $t = \text{TREF}$.

Note that, correspondingly, in static case $f(t) = 1.0$.

Barrier

Definitions

XBARRIER(1)	The X-coordinate of the first point of the barrier	$\tilde{X}(1)$
YBARRIER(1)	The Y-coordinate of the first point of the barrier	$\tilde{Y}(1)$
XBARRIER(2)	The X-coordinate of the second point of the barrier	$\tilde{X}(2)$
YBARRIER(2)	The Y-coordinate of the second point of the barrier	$\tilde{Y}(2)$

Remarks

A crack is modeled as a straightline connecting $\{\tilde{X}(1), \tilde{Y}(1)\}$ and $\{\tilde{X}(2), \tilde{Y}(2)\}$. For further development, a propagating crack may be modeled as a curve connecting a series of points $\{\tilde{X}(1), \tilde{X}(2), \ldots, \tilde{X}(n)\}$.

Printing Parameters

Definitions

MUPRINT	Number of points of which the displacements as functions of time can be plotted using EXCEL, MUPRINT < 5
MSPRINT	Number of elements of which the stresses as functions of time can be plotted using EXCEL, MSPRINT < 4
IUPRINT(I)	The Ith number of the point of which displacements are stored as functions of time
ISPRINT(I)	The Ith number of the element of which stresses are stored as functions of time

Remarks

1. For example, MUPRINT = 3 and IUPRINT(I = 1,2,3) = 2, 5, 8 means the displacements, u_x and u_y, of the second, the fifth, and the eighth node can be plotted as functions of time, using EXCEL.
2. For example, MSPRINT = 3 and ISPRINT(I = 1,2,3) = 3, 6, 9 means the stresses, σ_{xx}, σ_{yy}, and σ_{xy}, of the third, sixth, and ninth element (in the background mesh) can be plotted as functions of time, using EXCEL.

Appendix E
User's Manual: Meshless Computer Program for Analysis of Crack Growth in Elastoplastic Continuum

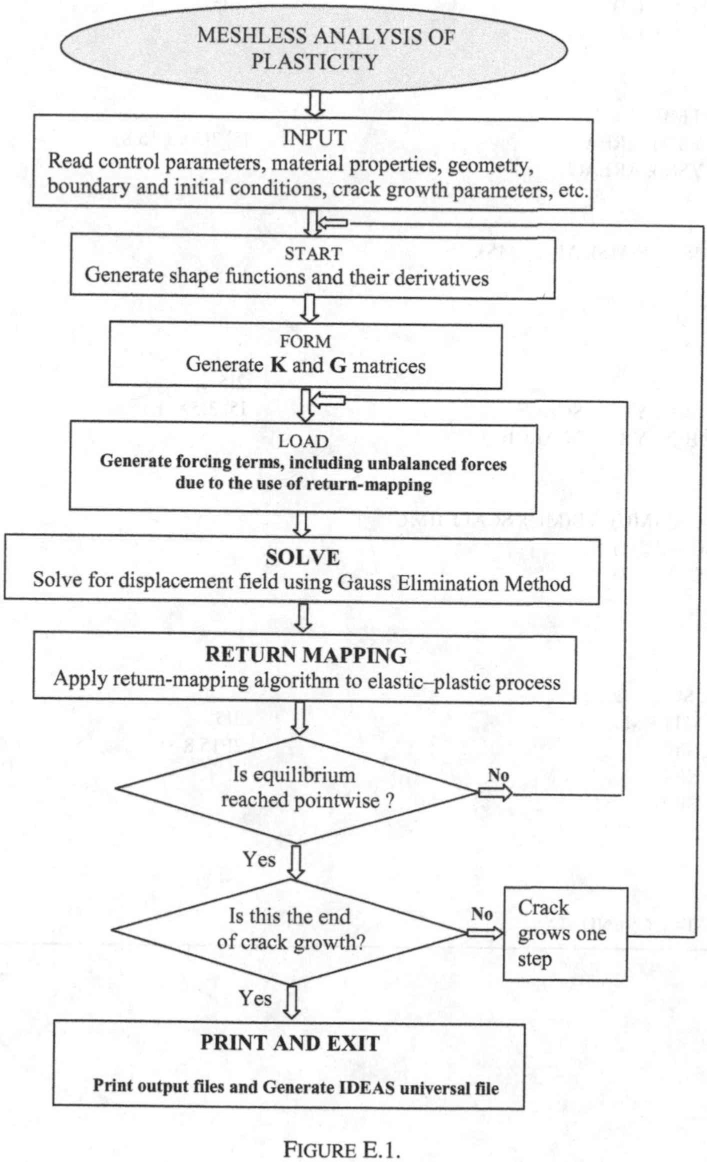

FIGURE E.1.

TABLE E.1 Table of input data (INFILE).

Variables	Format	Section
ID(1), ID(2), ..., ID(7)	16I5	1
MP, ME, MN, MS		
1, XNODE(1), YNODE(1), RNODE(1)	9X, I8, 7X, 3G8.5	2
2, XNODE(2), YNODE(2), RNODE(2)		
□		
□		
MN, XNODE(MN), YNODE(MN), RNODE(MN)		
1, (IJK(J,1), J = 1,4)	16I5	3
2, (IJK(J,2), J = 1,4)		
□		
□		
ME, (IJK(J,ME), J = 1,4)		
1, XS(1), YS(1), AREA(1)	I5, 3(5X,E15.8)	4
2, XS(2), YS(2), AREA(2)		
□		
□		
MS, XS(MS), YS(MS), AREA(MS)		
YOUNG	E15.8	5
POISSON		
VON		
PSLOPE		
MU, MF	2I5	6
ICU(1), XB(1), YB(1), SCALE1(1)	I5, 3(5X, E15.8)	
ICU(2), XB(2), YB(2), SCALE1(2)		
□		
□		
ICU(MU), XB(MU), YB(MU), SCALE1(MU)		
ICF(1), SCALE2(1)		
ICF(2), SCALE2(2)		
⋮		
⋮		
ICF(MF), SCALE2(MF)		
NDATA, MTERM	2I5	7
ASI(0), CSI(0)	2F15.8	
ASI(1), CSI(1)		
ASI(2), CSI(2)		
⋮		
⋮		
ASI(NDATA), CSI(NDATA)		

Control Parameters

Definitions

ID(1) = 0	Static case
ID(2) = 0	Local theory
ID(3) = 2	Second-order polynomial used in the shape function
ID(3) = 3	Third-order polynomial used in the shape function
ID(4) = 1	Plane stress
ID(5) = 0	There is no barrier
ID(5) = 1	There is a barrier
ID(6) = MSTEP	Number of load steps
ID(7) = 0	There is no mirror symmetry
ID(7) = 1	There is a mirror symmetry about x-axis
ID(7) = 2	There are mirror symmetries about x-axis and y-axis
MP	Number of points in the background mesh only for the use of IDEAS
ME	Number of elements in the background mesh only for the use of IDEAS
MN	Number of nodes (in this version MN = ME)
MS	Number of sampling points

Remarks

1. This is the User's Manual of a meshless computer program for the analysis of crack growth based on two-dimensional plane stress plasticity.
2. The line crack is assumed to be along the x-axis; the problem is assumed to have mirror symmetry about x-axis. For center-cracked problems, one may also assume there is additional mirror symmetry about the y-axis.
3. This computer program generates an IDEAS universal file, as part of the output, in order to graphically show the stress distributions in the whole specimen. For this reason, a finite element type background mesh is needed.
4. A crack, in meshless method, is considered as a barrier that may block the support of a node to a sampling point if the crack cuts through the straightline connecting these two points.
5. A sampling point is a point of which the detailed information is needed in the process of analysis. It is equivalent to a Gauss point in finite element analysis.
6. The second-order polynomial basis in two-dimensional domain is

$$\mathbf{p} = \{1, x, y, x^2, y^2, xy\}'$$

while the third-order polynomial basis is

$$\mathbf{p} = \{1, x, y, x^2, y^2, xy, x^3, y^3, x^2y, xy^2\}'.$$

Nodes

Definitions

XNODE(I)	X-coordinate of the Ith node
YNODE(I)	Y-coordinate of the Ith node
RNODE(I)	Radius of support of the Ith node

Remarks

1. Every point in the domain must be supported by at least n nodes; n is the number of terms in the polynomial basis.
2. The weight function between the sampling point $\mathbf{x}$ and the Ith node $\mathbf{x}_I$ is a quartic spline expressed as

$$w(\mathbf{x}, \mathbf{x}_I) \equiv w_I(\mathbf{x}) = \begin{cases} 1 - 6s^2 + 8s^3 - 3s^4, & \text{if } s \leq 1 \\ 0, & \text{if } s \geq 1 \end{cases}$$

where R_I is the radius of support of the Ith node and

$$s = \frac{||\mathbf{x} - \mathbf{x}_I||}{R_I}.$$

Background Mesh

Definitions

IJK(J,I)	The number of the Jth point of the Ith element, i.e., the connectivity of background mesh

Remarks

The background mesh is for the purpose of generating an IDEAS universal file; there are four points per element; the numbering of the four points goes counter-clockwise.

Sampling Points

Definitions

XS(I)	X-coordinate of the Ith sampling point
YS(I)	Y-coordinate of the Ith sampling point
AREA(I)	The area associated with the Ith sampling point

Remarks

1. The sampling points are equivalent to the Gauss points in finite element method.
2. Each sampling point is associated with an area; the summation of all the areas of all the sampling points should be equal to the total area of the specimen.
3. The areas of all sampling points should be supplied by the user; note that in finite element method each Gauss point is also associated with an area; however, in that case, the area can be calculated because the geometry of each element is specified.

Material Properties

Definitions

YOUNG (E)	Young's modulus of the material
POISSON (υ)	Poisson's ratio of the material
VON (σ_Y)	The von Mises yield strength
PSLOPE (H)	The slope of the incremental stress–strain relation in the plastic region

Remarks

1. The stress–elastic strain relation is expressed as

$$\sigma_{ij} = \lambda\left(\varepsilon_{kk} - \varepsilon^{\mathrm{p}}_{kk}\right)\delta_{ij} + 2\mu\left(\varepsilon_{ij} - \varepsilon^{\mathrm{p}}_{ij}\right);$$

and the two lame constants, λ and μ, are related to Young's modulus and Poisson's ratio.

2. The J_2 flow theory for plane stress plasticity is adopted in this code; σ_Y is the von Mises yield strength; and H is the slope of the incremental stress–strain relation in the plastic region (cf. Simo and Hughes, 1998).

Boundary Conditions

Definitions

MU	Number of displacement-specified boundary conditions
MF	Number of nodal-force-specified boundary conditions
ICU(I)	The X- or Y-component of the Ith-displacement-specified boundary condition
XB(I)	The X-coordinate of the location at which the Ith boundary condition is specified
YB(I)	The Y-coordinate of the location at which the Ith boundary condition is specified

SCALE1(I)	The magnitude of the displacement specified for the ICU(I)th component
ICF(I)	The component number of the Ith nodal force specified boundary condition
SCALE2(I)	The magnitude of the nodal force specified for the ICF(I)th component

Remarks

1. To specify the essential boundary conditions, take the following as an example. Let ICU(3) = 1, XB(3) = 1.01, YB(3) = 2.05, SCALE1(3) = 10.66. It means the third displacement-specified boundary condition should read as

$$U_x(1.01, 2.05) = 10.66.$$

In this example, $\{x, y\} = \{1.01, 2.05\}$ should be a point on the boundary, but it does not have to be a node. On the other hand, even if that point is a node, the nodal value is not equal to the value of displacement at that node. This is one of the key distinctions that separate finite element method and meshless method.

2. To specify the natural boundary condition, there is no difference between finite element method and meshless method. For example, if, as the mth natural boundary condition, we want to specify the X-component of the applied force at the 15th node, which is on the boundary, to be 1.2345, i.e.,

$$F_x(15) = 1.2345,$$

then we set ICF(m) = 29, SCALE2(m) = 1.2345.

3. In slow-crack growth analysis, the line crack is along the x-axis from $\{x, y\} = \{0, 0\}$ to $\{x, y\} = \{a^0, 0\}$ initially. Then the essential boundary conditions to represent the crack growth are specified as follows. First, for the initial crack tip

$$\begin{aligned} \text{ICU(MU)} &= 2, \\ \text{XB(MU)} &= a^0, \\ \text{YB(MU)} &= 0.0, \\ \text{SCALE1(MU)} &= 0.0. \end{aligned}$$

In an immediate next step, the crack grows to $\{x, y\} = \{a^1, 0\}$, then we have the following to represent the boundary condition at the crack tip

$$\begin{aligned} \text{ICU(MU} - 1) &= 2, \\ \text{XB(MU} - 1) &= a^1, \\ \text{YB(MU} - 1) &= 0.0, \\ \text{SCALE1(MU} - 1) &= 0.0. \end{aligned}$$

The order is important. For the sake of discussion, in the next step the crack grows to

$$\{x, y\} = \{a^2, 0\}$$

then the boundary condition at this crack tip is specified as

$$\begin{aligned} \text{ICU(MU} - 2) &= 2, \\ \text{XB(MU} - 2) &= a^2, \\ \text{YB(MU} - 2) &= 0.0, \\ \text{SCALE1(MU} - 2) &= 0.0. \end{aligned}$$

Crack Growth DATA

Definitions

NDATA	Number of data points in the applied stress–crack size curve (experimental data)
MTERM	Number of terms in the polynomial to represent the applied stress–crack size curve
ASI(0)	The applied stress at the onset of slow-crack growth $= \sigma_{yy}^0$
CSI(0)	The initial crack size $= a^0$
ASI(I)	The applied stress at the Ith data point $= \sigma_{yy}(I)$
CSI(I)	The crack size at the Ith data point $= a(I)$
ASI(NDATA)	The applied stress at the onset of fast fracture
CSI(NDATA)	The crack size at the onset of fast fracture

Remarks

1. The crack growth data is an experimental curve relating applied stress σ_{yy} and crack size a (a variable); it should be realistic, otherwise, the output will be misleading.
2. Suppose the data consist of 10 pairs of $\{\sigma_{yy}, a\}$, including the initial crack size a^0 and the applied stress at the onset of slow-crack growth σ_{yy}^0. Then NDATA $= 9(= 10 - 1)$ and MTERM ≤ 9. It is suggested to set MTERM $=$ NDATA.

Bibliography

Chapter 1

Cochran, W. (1973) *The Dynamics of Atoms in Crystals*, Edward Arnold Limited, London.

Devonshire, A. F. (1954) "Theory of ferroelectrics," *Phil. Mag. Suppl.* (*Adv. Phys.*) **3**(10), 85–130.

Ginzburg, V. L. (2001) "Phase transitions in ferroelectrics: Some historical remarks," in *The 10th International Conference on Ferroelectricity, Phys.-Usp.* **44**(10), 1037–1043.

Haile, J. M. (1992) *Molecular Dynamics Simulation*, John Wiley & Sons, New York.

Hohenberg, H. and Kohn, W. (1964) "Inhomogeneous gas," *Phys. Rev.* **136**, B864–B871.

Hoover, W. G. (1986) *Molecular Dynamics*, Springer-Verlag, Berlin.

Hoover, W. G. (1991) *Computational Statistical Mechanics*, Elsevier, Amsterdam.

Huang, K. (1967) *Statistical Mechanics*, John Wiley & Sons, New York.

Kittel, C. (1967) *Introduction to Solid State Physics*, John Wiley & Sons, New York.

Kohn, W. and Sham, L. J. (1965) "Self-consistant equations including exchange and correlation effect," *Phys. Rev.* **140**, A1133–A1138.

Parr, R. G. and Yang, W. (1989) *Density Functional Theory of Atoms and Molecules*, Oxford University Press, New York.

Tolman, R. C. (1962) *The Principle of Statistical Mechanics*, Oxford University Press, London.

Yu, P. Y. and Cardona, M. (2001) *Fundamentals of Semiconductors*, Springer-Verlag, Berlin.

Chapter 2

Collman, B. D. and Noll, W. (1963) "The thermodynamics of elastic materials with heat conduction," *Arch. Ration. Mech. Anal.* **13**, 167.

Eringen, A. C. (1989) *Mechanics of Continua*, Robert E. Krieger Publishing Company, Melbourne.

Eringen, A. C. (1999) *Microcontinuum Field Theories I: Foundations and Solids*, Springer-Verlag, New York.

Fung, Y. C. and Tong, P. (2001) *Classical and Computational Solid Mechanics*, World Scientific Publishing Company, Singapore.

Truesdell, C. and Noll, W. (1965) *The non-linear field theories of mechanics*, Handbuck der Physik , ed. S. Flugge, Vol. III/3 Springer-Verlag, Berlin.

Truesdell, C. and Toupin, R. (1960) *The classical field theory*, Handbuck der Physik , ed. S. Flugge, Vol. III/1 Springer-Verlag, Berlin.
Wang, C. C. (1970) "A new representation theorem for isotropic functions, Part I and Part II," *Arch. Ration. Mech. Anal.* **36**, 166–223.
Wang, C. C. (1971) "Corrigendum to representations for isotropic functions," *Arch. Ration. Mech. Anal.* **43**, 392–395.

Chapter 3

Bathe, K. J. (1996) *Finite Element Procedures*, Prentice-Hall, Englewood Cliffs.
Belytschko, T., Liu, W. K. and Moran, B. (2001) *Nonlinear Finite Elements for Continua and Structures*, John Wiley & Sons, Chichester, England.
Cook, R. D., Malkus, D. S. and Plesha, M. E. (1989) *Concepts and Applications of Finite Element Analysis*, John Wiley & Sons, New York.
Fung, Y. C. and Tong, P. (2001) *Classical and Computational Solid Mechanics*, World Scientific Publishing Company, Singapore.
Grandin, H., Jr. (1991) *Fundamentals of the Finite Element Method*, Waveland Press, Prospect Heights.
Logan, D. L. (1992) *A First Course in the Finite Element Method*, PWS-KENT Publishing Company, Boston.
Reddy, J. N. (1993) *An Introduction to the Finite Element Method*, McGraw-Hill, Boston.
Washizu, K. (1975) *Variational Methods in Elasticity and Plasticity*, Pergamon Press, Oxford.
Zienkiewicz, O. Z. (1983) *The Finite Element Method*, Mc-Graw-Hill, London.
Zienkiewicz, O. C. and Taylor, R. L. (1989) *The Finite Element Method: Volume 1: Basic Formulation and Linear Problems*, McGraw-Hill, London.
Zienkiewicz, O. C. and Taylor, R. L. (1991) *The Finite Element Method: Volume 2: Solid and Fluid Mechanics Dynamics and Non-linearity*, McGraw-Hill, London.

Chapter 4

Aifantis, E. C. (1999) "Strain gradient interpretation of size effect," *Int. J. Fract.* **95**, 299–314.
Aluru, N. R. (1999) "A reproducing kernel particle method for meshless analysis of micro-electromechanical systems," *Comput. Mech.* **23**, 324–338.
Atluri, S. N. and Zhu, T. (2000) "New concept in meshless methods," *Int. J. Numer. Methods Eng.* **47**, 537–556.
Attaway, S. W., Heimsten, M. W. and Swegle, J. W. (1994) "Coupling of smooth particle hydrodynamics with the finite element method," *Nucl. Eng. Des.* **150**, 199–205.
Babuska, I. and Melenk, J. M. (1995) "The partition of unity finite element method," Technique Report BN-1185, Institute for Physics, Science, and Technology, University of Maryland, Maryland.
Babuska, I. and Melenk, J. M. (1996) "The partition of unity finite element method: Basic theory and applications," *Comput. Methods Appl. Mech. Eng.* **139**, 289–315.
Beissel, S. and Belytschko, T. (1996) "Nodal integration of the element-free Galerkin method," *Comput. Methods Appl. Mech. Eng.* **139**, 49–74.

Belytschko, T., Lu, Y. Y. and Gu, L. (1994) "Element-free Galerkin methods," *Int. J. Numer. Methods Eng.* **37**, 229–256.

Belytschko, T., Organ, D. and Krongauz, Y. (1995) "A coupled finite element—Element free Galerkin method," *Comput. Mech.* **17**, 186–195.

Belytschko, T., Krongauz, Y., Organ, D., Fleming, M. and Krysl, P. (1996a) "Meshless methods: An overview and recent developments," *Comput. Methods Appl. Mech. Eng.* **139**,3–47.

Belytschko, T., Krongauz, Y., Fleming, M., Organ, D. and Liu, W. K. (1996b) "Smoothing and accelerated computations in the element free Galerkin method," *J. Comput. Appl. Math.* **74**, 111–126.

Belytschko, T., Krongauz, Y., Organ, D., Fleming, M. and Krysl, P. (1996c) "Meshless methods: An overview and recent developments," *Comput. Methods Appl. Mech. Eng.* **139**, 3–47.

Belytschko, T., Krysl, P. and Krongauz, Y. (1997) "A three-dimensional explicit element-free Galerkin method," *Int. J. Numer. Methods Fluid* **24**, 1253–1270.

Belytschko, T., Guo, Y., Liu, W. K. and Xiao, S. (2000) "A unified stability analysis of meshless particle methods," *Int. J. Numer. Methods Eng.* **48**, 1359–1400.

Braun, J. and Sambridge, M. (1995) "A numerical method for solving partial differential equations on highly irregular evolving grids," *Nature* **376**,655–660.

Chen, J., Pan, C., Wu, C. and Liu, W. K. (1996) "Reproducing kernel particle methods for large deformation analysis of non-linear structures," *Comput. Methods Appl. Mech. Eng.* **139**, 195–227.

Chen, J., Wu, C., Yoon, S. and You, Y. (2001a) "A stabilized conforming nodal integration for Galerkin mesh-free methods," *Int. J. Numer. Methods Eng.* **50**, 435–466.

Chen, Y., Lee, J. D. and Eskandarian, A. (2001b) "Meshless particle method for nonlocal continua," in *Proceedings of International Conference of Computational Engineering Science*, Puerto Vallarta, Mexico.

Chen, J., Wu, C. and Belytschko, T. (2002a) "Regularization of material instabilities by meshfree approximations with intrinsic length scales," *Int. J. Numer. Methods Eng.* **47**, 1303–1322.

Chen, Y., Lee, J. D. and Eskandarian, A. (2002b) "Dynamic meshless method applied to nonlocal cracked problems," *Theor. Appl. Fract. Mech.* **38**, 293–300.

Chu, Y. and Moran, B. (1995) "A computational model for nucleation of solid–solid phase transformation," *Model. Simul. Mater. Sci. Eng.* **3**, 455–471.

Cueto, E., Doblare, M. and Gracia, L. (2000) "Imposing essential boundary conditions in natural elment method by means of density-scaled α-shapes," *Int. J. Numer. Methods Eng.* **48**, 519–546.

Donning, B. M. and Liu, W. K. (1998) "Meshless methods for shear-deformable beams and plates," *Comput. Methods Appl. Mech. Eng.* **152**, 47–71.

Duarte, C. A. M. and Oden, J. T. (1995) "Hp clouds—A meshless method to solve boundary value problems," TICAM Report, 95-05.

Duarte, C. A. M. and Oden, J. T. (1996) "An h-p adaptive method using clouds," *Comput. Methods Appl. Mech. Eng.* **139**, 237–262.

Dyka, C. (1994) "Addressing tensile instability in SPH methods," Technical Report NRL/MR/6384, NRL.

Dyka, C., Randles, P. and Ingel, R. (1997) "Stress points for tension instability in SPH," *Int. J. Numer. Methods Eng.* **40**, 2325–2341.

Edelen, D. G. B. (1969) "Protoelastic bodies with large deformation," *Arch. Ration. Mech. Anal.* **34**, 283.

Edelen, D. G. B. and Laws, N. (1971) "On the thermodynamics of systems with nonlocality," *Arch. Ration. Mech. Anal.* **43**, 24–35.

Eringen, A. C. (1966) "A unified theory of thermomechnical materials," *Int. J. Eng. Sci.* **4**, 179.

Eringen, A. C. and Edelen, D. G. B. (1972) "On nonlocal elasticity," *Int. J. Eng. Sci.* **10**, 233–248.

Fleck, N. A., Muller, G. M., Ashby, M. F. and Hutchinson, J. W. (1994) "Strain gradient plasticity: Theory and experiment," *Acta Metall. Mater.* **42**, 475–487.

Gingold, R. A. and Monaghan, J. J. (1977) "Smoothed particle hydrodynamics: Theory and application to non-spherical stars," *Mon. Not. R. Astron. Soc.* **181**, 375–387.

Grindeanu, I., Chang, K. H., Choi, K. K. and Chen, J. (1998) "Design sensitivity analysis of Hyperelastic structures using a meshless method," *AIAA J.* **36**, 618–627.

Grindeanu, I., Choi, K. K. and Chen, J. (1999) "Shape design optimization of hyperelastic structures using a meshless method," *AIAA J.* **37**, 990–997.

Hoover, W. G. and Hoover, C. G. (1993) "Smoothed-particle hydrodynamics and nonequalibrium molecular dynamics", Advanced Computational Methods for Material Modeling, AMD-Vol. 180/PVP-Vol. 268, ASME, New York, 231-242.

Hulbert, G. M. (1996) "Application of reproducing kernel particle methods in electromagnetics," *Comput. Methods Appl. Mech. Eng.* **139**, 229–235.

Johnson, G. R. (1994) "Linking Lagrangian particle method to standard finite element method for high velocity impact computations," *Nucl. Eng. Des.* **150**, 265–274.

Johnson, G. R. and Beissel, S. R. (1996) "Normalized smoothing functions for SPH impact computations," *Int. J. Numer. Methods Eng.* **39**, 2725–2741.

Jun, S., Liu, W. K. and Belytschko, T. (1998) "Explicit reproducing kernel particle methods for large deformation problems," *Int. J. Numer. Methods Eng.* **41**, 137–166.

Kim, N. H., Choi, K. K., Chen, J. S. and Park, Y. H. (2000) "Meshless shape design sensitivity analysis and optimization for contact problem with friction," *Comput. Mech.* **25**, 157–168.

Krongauz, Y. and Belytschko, T. (1997) "Consistent pseudo-derivatives in meshless methods," *Comput. Methods Appl. Mech. Eng.* **146**, 371–386.

Krongauz, Y. and Belytschko, T. (1998) "EFG approximation with discontinuous derivatives," *Int. J. Numer. Methods Eng.* **41**, 1215–1233.

Krysl, P. and Belytschko, T. (1996a) "Analysis of thin shells by the element-free Galerkin method," *Int. J. Solids Struct.* **33**, 3057–3080.

Krysl, P. and Belytschko, T. (1996b) "Analysis of thin palates by element free Galerkin methods, plates," *Comput. Mech.* **17**, 26–35.

Krysl, P. and Belytschko, T. (1996c) "Analysis of thin shells by element free Galerkin methods, plates," *Int. J. Solids Struct.* **33**, 3057–3080.

Krysl, P. and Belytschko, T. (1997) "Element-free Galerkin method: Convergence of the continuous and discontinuous shape functions," *Comput. Methods Appl. Mech. Eng.* **148**, 257–277.

Krysl, P. and Belytschko, T. (1999) "The element free Galerkin method for dynamic propagation of arbitrary 3-D cracks," *Int. J. Numer. Methods Eng.* **44**, 767–800.

Krysl, P. and Belytschko, T. (2000) "An efficient linear-precision partition of unity basis for unstructured meshless methods," *Commun. Numer. Methods Eng.* **16**, 239–255.

Lancaster, P. and Salkauskas, K. (1981) "Surface generated by moving least squares methods," *Math. Comput.* **37**, 141–158.

Li, S. and Liu, W. K. (1996) "Moving least-square reproducing kernel method Part II: Fourier analysis," *Comput. Methods Appl. Mech. Eng.* **139**, 159–193.

Li, S. and Liu, W. K. (2000) "Numerical simulations of strain localization in inelastic solids using mesh-free methods," *Int. J. Numer. Methods Eng.* **48**, 1285–1309.

Li, S., Hao, W. and Liu, W. K. (2000) "Numerical simulations of large deformation of thin shell structures using meshfree methods," *Comput. Mech.* **25**, 102–116.

Liszka, T. and Orkisz, J. (1980) "The finite difference method for arbitrary meshes," *Comput. Struct.* **5**, 45–58.

Liszka, T., Duarte, C. A. M. and Tworzydlo, W. W. (1996) "*hp*-Meshless cloud method," *Comput. Methods Appl. Mech. Eng.* **139**, 263–288.

Liu, W. K. and Chen, Y. (1995) "Wavelet and multiple scale reproducing kernel methods," *Int. J. Numer. Methods Fluid* **21**, 901–931.

Liu, W. K., Chen, Y. and Uras, R. A. (1995a) "Enrichment of the finite element method with the reproducing kernel particle method", Current Topics in Computational Mechanics, ed. J. F. Cory, Jr. and J. L. Gordon, PVP-Vol. 305, ASME, New York, 253–258.

Liu, W. K., Jun, S. and Zhang, Y. F. (1995b) "Reproducing kernel particle methods," *Int. J. Numer. Methods Fluid* **20**, 1081–1106.

Liu, W. K., Chen, Y., Chang, C. T. and Belytschko, T. (1996a) "Advances in multiple scale kernel particle methods," *Comput. Mech.* **18**, 73–111.

Liu, W. K., Chen, Y., Uras, R. A. and Chang, C. T. (1996b) "Generalized multiple scale reproducing kernel particle methods," *Comput. Methods Appl. Mech. Eng.* **139**, 91–157.

Liu, W. K., Li, S. and Belytschko, T. (1997) "Moving least-square reproducing kernel methods (I) Methodology and convergence," *Comput. Methods Appl. Mech. Eng.* **143**, 113–154.

Liu, W. K., Hao, S., Belytschko, T., Li, S. and Chang, C. (2000) "Multi-scale methods," *Int. J. Numer. Methods Eng.* **47**, 1343–1361.

Lu, Y. Y., Belytschko, T. and Gu, L. (1994) "A new implementation of the element free Galerkin," *Comput. Methods Appl. Mech. Eng.* **113**, 397–414.

Lucy, L. B. (1977) "A numerical approach to the testing of the fission hypothesis," *Astron. J.* **8**(12), 1013–1024.

Monaghan, J. J. (1982) "Why particle methods work," *SIAM J. Sci. Stat. Comput.* **3**(4), 422.

Monaghan, J. J. (1988) "An introduction to SPH," *Comput. Phys. Commun.* **48**, 89–96.

Monaghan, J. J. (1992) "Smooth particle hydrodynamics," *Annu. Rev. Astron. Astrophys.* **30**, 543–574.

Nayroles, B., Touzot, G. and Villon, P. (1992) "Generalizing the finite element method: Diffuse approximation and diffuse elements," *Comput. Mech.* **10**, 307–318.

Oden, J. T., Duarte, C. A. M. and Zienkiewicz, O. C. (1998) "A new cloud-based *hp* finite element method," *Comput. Methods Appl. Mech. Eng.* **153**, 117–126.

Onate, E., Idelsohn, S., Zienkiewicz, O. C. and Taylor, R. L. (1996a) "Finite point method in computational mechanics. Applications to convective transport and fluid flow," *Int. J. Numer. Methods Eng.* **39**, 3839–3866.

Onate, E., Idelsohn, S., Zienkiewicz, O. C., Taylor, R. L. and Sacco, C. (1996b) "Stabilized finite point method for analysis of fluid mechanics problems," *Comput. Methods Appl. Mech. Eng.* **139**, 315–346.

Ponthot, J. P. and Belytschko, T. (1998) "Arbitrary Lagrangian–Eulerian formulation for element-free Galerkin method," *Comput. Methods Appl. Mech. Eng.* **152**, 19–46.

Qian, S. and Weiss, J. (1993) "Wavelet and the numerical solution of partial differential equations," *J. Comput. Phys.* **106**, 155.

Randles, P. and Libersky, L. (1996) "Smoothed particle hydrodynamics: Some recent improvements and applications," *Comput. Methods Appl. Mech. Eng.* **139**, 375–408.

Randles, P., Libersky, L. and Petschek, A. (1999) "On neighbors, derivatives, and viscosity in particle codes," in *Proceedings of ECCM Conference*, Munich, Germany, 31 August to 3 September 1999.

Schwer, L. E., Gerlach, C. and Belystchko, T. (2000) "Element-free Galerkin simulations of concrete failure in dynamic uniaxial tension test", *J. Eng. Mech.* **126**(5), 443-454.

Senturia, S., Aluru, N. and White, J. (1997) "Simulating the behavior of MEMS devices: Computational methods and needs," *IEEE Comput. Sci. Eng.* **4**(1), 30–43.

Sibson, R. (1981) "A brief description of natural neighbor interpolationts," in *Interpreting Multivariate Data*, ed. V. Barnett, John Wiley & Sons, Chichester, England, pp. 21–36.

Sukumar, N. and Moran, B. (July, 1999) "C^1 natural neighbor interpolant for partial differential equations," *Numer. Methods Partial Differ. Equations* **15**(4), 417–447.

Sukumar, N., Moran, B. and Belytschko, T. (1998) "The natural element method in solid mechanics," *Int. J. Numer. Methods Eng.* **43**, 839–887.

Sukumar, N., Moran, B., Semenov, Y. and Belikov, V. V. (2001) "Natural neighbor Galerkin methods," *Int. J. Numer. Methods Eng.* **50**(1), 1–27.

Sulsky, D., Chen, Z. and Schhreyer, H. L. (1992) "The application of a material-spatial numerical method to penetration", New Methods in Transient Analysis, eds. P. Smolinski, W. K. Liu, G. Hulbert and K. Tamma, AMD-Vol. 143/PVP-Vol. 246, ASME, New York, 91–102.

Swegle, J. W., Hicks, D. L. and Attaway, S. W. (1995) "Smoothed particle hydrodynamics stability analysis," *J. Comput. Phys.* **116**, 123–134.

Traversoni, L. (1994) "Natural neighbor finite elements", *International Conference on Hydraulic Engineering Software, Hydrosoft Proceedings*, Vol. 2, 291–297, Computational Mechanics Publications, London.

Wagner, G. J. and Liu, W. K. (2000) "Application of essential boundary condition in mesh-free methods: A corrected collocation method," *Int. J. Numer. Methods Eng.* **47**, 1367–1379.

Yagawa, G. and Furukawa, T. (2000) "Recent development of free mesh method," *Int. J. Numer. Methods Eng.* **47**, 1419–1443.

Yagawa, G. and Yamada, T. (1996) "A new meshless finite element method," *Int. J. Comput. Mech.* **18**, 383–386.

Zhu, T. (1999) "A new meshless regular local boundary integral equation (MRLBIE) approach," *Int. J. Numer. Methods Eng.* **46**, 1237–1252.

Chapter 5

Atluri, S. N. and Zhu, T. (2000) "New concept in meshless methods," *Int. J. Numer. Methods Eng.* **47**, 537–556.

Beissel, S. and Belytschko, T. (1996) "Nodal integration of the element-free Galerkin method," *Comput. Methods Appl. Mech. Eng.* **139**, 49–74.

Belytschko, T., Lu, Y. Y. and Gu, L. (1994) "Element-free Galerkin methods," *Int. J. Numer. Methods Eng.* **37**, 229–256.

Belytschko, T., Krongauz, Y., Organ, D., Fleming, M. and Krysl, P. (1996) "Meshless methods: An overview and recent developments," *Comput. Methods Appl. Mech. Eng.* **139**, 3–47.

Belytschko, T., Krysl, P. and Krongauz, Y. (1997) "A three-dimensional explicit element-free Galerkin method," *Int. J. Numer. Methods Fluid* **24**, 1253–1270.

Belytschko, T., Guo, Y., Liu, W. K. and Xiao, S. (2000) "A unified stability analysis of meshless particle methods," *Int. J. Numer. Methods Eng.* **48**, 1359–1400.

Chen, Y. P., Lee, J. D. and Eskandarian, A. (2002) "Dynamic meshless method applied to nonlocal crack problems," *Theor. Appl. Fract. Mech.* **38**, 293–300.

Krysl, P. and Belytschko, T. (1997) "Element-free Galerkin method: Convergence of the continuous and discontinuous shape functions," *Comput. Methods Appl. Mech. Eng.* **148**, 257–277.

Monaghan, J. J. (1992) "Smooth particle hydrodynamics," *Annu. Rev. Astron. Astrophys.* **30**, 543–574.

Chapter 6

Bathe, K. J. (1982) *Finite Element Procedures in Engineering Analysis*, Prentice-Hall, Englewood Cliffs.

Bathe, K. J. (1996) *Finite Element Procedures*, Prentice-Hall, Englewood Cliffs.

Belytschko, T., Liu, W. K. and Moran, B. (2001) *Nonlinear Finite Elements for Continua and Structures*, John Wiley & Sons, Chichester, England.

Casal, P. (1978) "Interpretation of the rice integral in continuum mechanics," *Lett. Appl. Eng. Sci.* **16**, 335–347.

Chen, Y. P. and Lee, J. D. (2005) "Material force for dynamic crack propagation in multiphase micromorphic materials", *Theor. Appl. Fract. Mech.* **43**, 335–341.

Eshelby, J. D. (1951) "The force on an elastic singularity", *Phil. Trans. R. Soc. Lond.* **A244**, 87–112.

Hestenes, M. R. and Stiefel, E. (1952) "Methods of conjugate gradients for solving linear systems," *J. Res. Natl Bureau Stand.* **49**, 409–436.

Houbolt, J. C. (1950) "A recurrence matrix solution for the dynamic response of elastic aircraft," *J. Aeronaut. Sci.* **17**, 540–550.

Lee, J. D., Chen, Y. and Yin, S. W. (2004) "Material forces in micromorphic fracture mechanics", *Journal of the Chinese Institute of Engineers* **27**(6), 889-896.

Lee, J. D. and Chen, Y. (2005) "Material forces in micromorphic thermoelastic solids", *Philosophical Magazine* **85**(33–35), 3897–3910.

Maugin, G. A. (1992a) *Thermomechanics of Plasticity and Fracture*, Cambridge University Press, United Kingdom.

Maugin, G. A. (1992b) "Application of an energy-momentum tensor in nonlinear elastodynamics: Pseudomomentum and Eshelby stress in solitonic elastic systems," *J. Mech. Phys. Solids* **40**, 1543–1558.

Maugin, G. A. (1993) *Material Inhomogeneities in Elasticity*, Chapman & Hall, London.

Maugin, G. A. (1995) "Material forces: Concepts and applications," *Appl. Mech. Rev.* **48**, 213–245.

Newmark, N. M. (1959) "A methods of computation for structural dynamics," *ASCE J. Eng. Mech.* **85**, 67–94.

Peach, M. O. and Koehler, J. S. (1950) "Forces exerted on dislocations and the stress field produced by them," *Phys. Rev.* **II-80**, 436–439.

Rice, J. R. (1968) "Path-independent integral and the approximate analysis of strain concentrations by notches and cracks," *Trans. ASME J. Appl. Mech.* **33**, 379–385.

Sih, G. C. and Liebowitz, H. (1968) "Mathematical theories of brittle fracture," in *Fracture II*, ed. H. Liebowitz, Academic Press, New York.

Sneddon, I. N. and Lowengrub, M. (1969) *Crack Problems in the Classical Theory of Elasticity*, John Wiley & Sons, New York.

Tong, P. and Rossettos, J. N. (1977) *Finite Element Method Basic Technique and Implementation*, The MIT Press, Cambridge.

Varga, R. S. (1962) *Matrix Iterative Analysis*, Prentice-Hall, Englewood Cliffs.

Wilson, E. L., Farhoomard, I. and Bathe, K. J. (1973) "Nonlinear dynamic analysis of complex structures," *Int. J. Earthq. Eng. Struct. Dyn.* **1**, 241–252.

Chapter 7

Aifantis, E. C. (1999) "Strain gradient interpretation of size effect," *Int. J. Fract.* **95**, 299–314.

Bazant, Z. P. and Lin, F. (1988) "Non-local yield limit degradation," *Int. J. Numer. Methods Eng.* **26**, 1805–1823.

Bazant, Z. P. and Pijaudier-Cabot, G. (1988) "Nonlocal continuum damage, localization instability and convergence," *J. Appl. Mech.* **55**, 287–293.

Belytschko, T., Lu, Y. Y. and Gu, L. (1994) "Element-free Galerkin methods," *Int. J. Numer. Methods Eng.* **37**, 229–256.

Belytschko, T., Krongauz, Y., Organ, D., Fleming, M., Krysl, P. (1996) "Meshless methods: An overview and recent developments," *Comput. Methods Appl. Mech. Eng.* **139**, 3–47.

Belytschko, T., Guo, Y., Liu, W. K. and Xiao, S. P. (2000) "A unified stability analysis of meshless particle methods", *Int. J. Numer. Methods Eng.* **48**, 1359–1400.

Charalambakis, N. and Aifantis, E. C. (1991) "Thermoviscoplastic shear instability and higher order strain gradients," *Int. J. Eng. Sci.* **29**, 1639–1650.

Chen, J., Wu, C. and Belytschko, T. (2000) "Regularization of material instability by meshfree approximations with intrinsic scales," *Int. J. Numer. Methods Eng.* **47**, 1303–1322.

Chen, Y. P., Lee, J. D. and Eskandarian, A. (2002) "Dynamic meshless method applied to nonlocal crack problems," *Theor. Appl. Fract. Mech.* **38**, 293–300.

Edelen, D. G. B. (1969) "Protoelastic bodies with large deformation," *Arch. Ration. Mech. Anal.* **34**, 283.

Edelen, D. G. B. and Laws, N. (1971) "On the thermodynamics of systems with nonlocality," *Arch. Ration. Mech. Anal.* **43**, 24–35.

Edelen, D. G. B., Green, A. E. and Laws, N. (1971) "Nonlocal continuum mechanics," *Arch. Ration. Mech. Anal.* **43**, 36–44.

Eringen, A. C. (1966) "A unified theory of thermomechanical materials," *Int. J. Eng. Sci.* **4**, 179.

Eringen, A. C. (1976) "Nonlocal polar field theories," in *Continuum Physics*, ed. A. C. Eringen, Academic Press, New York.

Eringen, A. C., Edelen, D. G. B. (1972) "On nonlocal elasticity," *Int. J. Eng. Sci.* **10**, 233–248.

Fleck, N. A. and Hutchinson, J. W. (1997) "Strain gradient plasticity," *Adv. Appl. Mech.* **13**, 295–361.

Fleck, N. A., Muller, G. M., Ashby, M. F. and Hutchinson, J. W. (1994) "Strain gradient plasticity: Theory and experiment," *Acta Metall. Mater.* **42**, 475–487.

Griffel, D. F. (1981) *Applied Functional Analysis*, John Wiley & Sons, New York.

Hoover, W. G. (1986) *Molecular Dynamics*, Springer-Verlag, Berlin Heidelberg.

Hoover, W. G. (1991) *Computational Statistical Mechanics*, Elsevier, Amsterdam; Oxford, New York, Tokyo.

Hoover, W. G. and Hoover, C. G. (1993) "Smoothed-particle hydrodynamics and nonequalibrium molecular dynamics", Advanced Computational Methods for Material Modeling, AMD-Vol. 180/PVP-Vol. 268, ASME, New York, 231–242.

Jun, S., Liu, W. K. and Belytschko, T. (1998) "Explicit reproducing kernel particle methods for large deformation problems," *Int. J. Numer. Methods Eng.* **41**, 137–166.

Maki-Jaskari, M. (2001) "Simulations of strain relief at the crack tip in silicon," *J. Phys. Condens. Matter* **13**, 1429–1437.

Masuda-Jindo, K., Tewary, V. K. and Thomson, R. (1991) "Atomic theory of fracture of brittle materials: Application to covalent semiconductors", *Journal of Materials Research* **6**, 1553–1566.

Nilsson, C. (1998) "On nonlocal rate-independent plasticity," *Int. J. Plast.* **14**(6), 551–575.

Rudd, R. E. and Broughton, J. Q. (1999) "Atomistic simulation of MEMS resonators through the coupling of length scales," *J. Model. Simul. Microsys.* **1**, 26.

Sih, G. C. and Liebowitz, H. (1968) "Mathematical theories of brittle fracture," in *Fracture II*, ed. H. Liebowitz, Academic Press, New York.

Timoshenko, S. and Goodlier, J. N. (1969) *Theory of Elasticity*, McGraw-Hill, New York.

Chapter 8

Belytschko, T., Lu, Y. T. and Gu, L. (1994) "Element-free Galerkin methods," *Int. J. Numer. Methods Eng.* **37**, 229–256.

Belytschko, T., Krongauz, Y., Organ, D., Fleming, M. and Krysl, P. (1996) "Meshless methods: An overview and recent developments," *Comput. Methods Appl. Mech. Eng.* **139**, 3–47.

Belytschko, T., Guo, Y., Liu, W. K. and Xiao, S. P. (2000a) "A unified stability analysis of meshless particle methods," *Int. J. Numer. Methods Eng.* **48**, 1359–1400.

Belytschko, T., Liu, W. K. and Moran, M. (2000b) *Nonlinear Finite Elements for Continua and Structures*, John Wiley & Sons, Chichester, England.

Casey, J. (1998) "On elastic-thermo-plastic materials at finite deformations," *Int. J. Plast.* **14**, 173–191.

Chen, Y., Eskandarian, A., Oskard, M. and Lee, J. D. (2004) "Meshless analysis of plasticity with application to crack growth problems", *Theor. Appl. Fract. Mech.* **41**, 83–94.

Chen, Y., Lee, J. D. and Eskandarian, A. (2002) "Dynamic meshless method applied to nonlocal cracked problems," *Theor. Appl. Fract. Mech.* **38**, 293–300.

Green, A. E. and Naghdi, P. M. (1965) "A general theory of an elastic-plastic continuum," *Arch. Ration. Mech. Anal.* **18**, 251–281.

Herstein, J. N. (1964) *Topics in Algebra*, Ginn and Company, Waltham.

Hill, R. (1950) *The Mathematical Theory of Plasticity*, Oxford University Press, Oxford, United Kingdom.

Hungerford, T. W. (1974) *Algebra*, Springer-Verlag, New York.

Johnson, C. (1978) "On plasticity with hardening," *J. Appl. Math. Anal.* **62**, 325–336.

Jun, S., Liu, W. K. and Belytschko, T. (1998) "Explicit reproducing kernel particle methods for large deformation problems," *Int. J. Numer. Methods Eng.* **41**, 137–166.

Krieg, R. D. and Key, S. W. (1976) "Implementation of a time dependent plasticity theory into structural computer programs," in *Constitutive Equations in Viscoplasticity:*

Computational and Engineering Aspects, eds. J. A. Stricklin and K. J. Saczalski AMD-20, ASME, New York.

Lee, J. D. and Chen, Y. (2001) "A theory of thermo-visco-elastic-plastic materials: Thermomechanical coupling in simple shear," *Theor. Appl. Fract. Mech.* **35**, 187–209.

Lee, J. D., Liebowitz, H. and Lee, K. Y. (1996) "The quest of a universal fracture law governing the process of slow crack growth," *Eng. Fract. Mech.* **55**, 61–83.

Lee, K. Y., Liebowitz, H. and Lee, J. D. (1997) "Finite element analysis of the slow crack growth process in mixed mode fracture," *Eng. Fract. Mech.* **56**, 551–577.

Lubliner, J. (1984) "A maximum-dissipation principle in generalized plasticity," *Acta Mech.* **52**, 225–237.

Lubliner, J. (1986) "Normality rules in large-deformation plasticity," *Mech. Mater.* **5**, 29–34.

Moreau, J. J. (1976) "Application of convex analysis to the treatment of elastoplastic systems," in *Applications of Methods of Functional Analysis to Problems in Mechanics*, eds. P. Germain and B. Nayroles, Springer-Verlag, Berlin.

Simo, J. C. and Hughes, T. R. (1998) *Computational Inelasticity*, Springer, New York.

Simo, J. C., Kennedy, J. G. and Govindjee, S. (1988) "Non-smooth multisurface plasticity and viscoplasticity: Loading/unloading conditions and numerical algorithms," *Int. J. Numer. Methods Eng.* **26**, 2161–2185.

Appendix

Boehler, J. P. (1977) "On irreducible representations for isotropic scalar functions," *ZAMM* **57**, 323–327.

Simo, J. C. and Hughes, T. R. (1998) *Computational Inelasticity*, Springer, New York.

Smith, G. F. (1970) "On a fundamental error in two papers of C. C. Wang 'on representations for isotropic functions, Part I and II'," *Arch. Ration. Mech. Anal.* **36**, 161–165.

Smith, G. F. (1971) "On isotropic functions of symmetric tensors, skew-symmetric tensors and vectors," *Int. J. Eng. Sci.* **9**, 899–916.

Wang, C. C. (1969a) "On representations for isotropic functions, Part I," *Arch. Ration. Mech. Anal.* **33**, 249.

Wang, C. C. (1969b) "On representations for isotropic functions, Part II," *Arch. Ration. Mech. Anal.* **33**, 268.

Wang, C. C. (1970) "A new representation theorem for isotropic functions, Part I and Part II," *Arch. Ration. Mech. Anal.* **36**, 166–223.

Wang, C. C. (1971) "Corrigendum to 'representations for isotropic functions'," *Arch. Ration. Mech. Anal.* **43**, 392–395.

Index